HYDROMETRY
THIRD EDITION

# UNESCO-IHE LECTURE NOTE SERIES

# Hydrometry

## 3rd edition

*A comprehensive introduction to the measurement of flow in open channels*

W. BOITEN

*Wageningen University, The Netherlands*
*UNESCO-IHE Institute for Water Education, Delft, The Netherlands*

CRC Press
Taylor & Francis Group
Boca Raton   London   New York   Leiden

CRC Press is an imprint of the
Taylor & Francis Group, an **informa** business

A BALKEMA BOOK

*CRC Press/Balkema is an imprint of the Taylor & Francis Group, an informa business*

© 2008 Taylor & Francis Group, London, UK

Typeset by Vikatan Publishing Solutions (P) Ltd., Chennai, India
Printed and bound in Great Britain by Antony Rowe (A CPI-group Company), Chippenham, Wiltshire.

Published by: CRC Press/Balkema
P.O. Box 447, 2300 AK Leiden, The Netherlands
e-mail: Pub.NL@taylorandfrancis.com
www.crcpress.com – www.taylorandfrancis.co.uk – www.balkema.nl

ISBN: 978-0-415-46763-6 (paperback edition)

# Contents

# Preface

On the Earth an increasing need exists for safe drinking water and proper sanitation. At present, almost 60% of the estimated annually available fresh water is being used. The water balance of a hydrological catchment provides the required information of available water amounts at a local scale. One of the water balance components is the discharge in open channels and rivers. In Hydrometry we discuss universally accepted methods for the measurement of flow in open water courses.

In this third edition more attention is paid to modern methods and developments, as far as standardized in the ISO standards on Hydrometry. The sections on International Standards have been updated to 2007.

Likewise the cover of the Book has been renewed. The photo of the survey vessel has been taken by Mr. Jan Tekstra, Rijkswaterstaat, the Dutch Directorate for Public Works and Water Management.

Wageningen, January 2008                                  Wubbo Boiten, author

# Preface to the Second Edition

The preparation of this second edition was assisted greatly by the constructive criticisms of colleagues and users of the text. The initial aims of *Hydrometry* to be an introduction of the subject to undergraduates, participants of graduate courses and professionals in the field of water resources have been realized.

The opportunity has been taken to include a section on fish migration which completes Chapter 6, Flow measurement structures.

In addition the sections on International Standards have been updated.

Wageningen, January 2003                                  Wubbo Boiten, author

# Preface to the First Edition

Hydrometry is the science of measuring flow in open channels including methods, techniques and instrumentation used in hydrology and water resources management. Measurements are fundamental to the development and application of hydrological science.

'Hydrometry' is addressed primarily to undergraduates having a basic knowledge of hydraulics and hydrology, but also to participants of graduate courses and professionals in the field of water resources.

It is based on lecturing material developed for undergraduates of Wageningen University, Department of Environmental Sciences. The first edition was issued in 1987 as lecture notes for students of the Sub-department Water Resources, and the Irrigation and Water Engineering Group.

The lecture series on 'Hydrometry' is furthermore part of various International Masters Programmes on Hydrology, Water Resources and Hydraulic Engineering, as well as short courses organized by the International Institute for Infrastructure, Hydraulic and Environmental Engineering (IHE), Delft.

The present second edition of 'Hydrometry' is based on many years of lecturing experience with Dutch students and participants of Masters programmes in Delft from many countries. The notes have now been updated to meet the latest requirements of professionals in the field of Water Resources Engineering.

The main topics include: measurements of water levels and bed levels, discharge measurements (three methods for single measurements and six methods for continuous measurements), measurement of sediment transport, use of flow measurement structures, hydrological networks, and the organization of surveys.

Knowledge of hydrometric determinations is important in various fields of water engineering, such as:
- daily water management of rural and urban areas, including the water distribution in irrigation schemes,
- design of reservoirs, water supply systems, bridges and culverts,
- performance of bilateral agreements on the distribution of river water,
- forecast and management of floods,
- water balance studies.

The hydrometry in this book deals only with:
- land surface, not coastal or marine areas (hydrography)
- water quantity, not water quality, and
- pure measurement techniques, not statistical elaborations.

This book is intended to be a contribution towards a better hydrometric knowledge and practice. It deals with both traditional techniques and new methods such as Global Positioning System (GPS), the Acoustic Doppler Current Profiler (ADCP), and others. The majority of the methods and the instruments are in accordance with the latest ISO standards.

I wish to acknowledge my colleagues of Wageningen University, WL Delft Hydraulics, and IHE Delft who supported and encouraged me. I am especially grateful to the team of colleagues for their contribution to the format of this book: Mr. Rudolf Smit, Mr. Piet Kostense and Mrs. Henny van Werven.

The reviewers of this book are also gratefully acknowledged: Mr A. Dommerholt, Wageningen University; Mr P. van Groen, WL Delft Hydraulics; Mr M. Huygens, the State University of Ghent; Mr P.J.M. de Laat, IHE Delft.

I hope that 'Hydrometry' will find its way to all students and professionals who are engaged in the management of Water and Environmental Resources. Any comments which may lead to a broader insight into flow measurement practices will be welcomed.

Wageningen, April 2000                                    Wubbo Boiten, author

CHAPTER 1

# Introduction

Hydrometry means literally water measurement. In the past hydrometric engineers were particularly involved in streamflow measurements. Today many aspects of water measurements are included.
Hydrometry is defined in this course as *the measurement of flow in open watercourses, supported or complemented by the measurements of water levels, bed levels and sediment transport.*

Flow measurement (discharge measurement) constitutes the principal part in hydrometry as shown in the diagramme below.

| Supporting measurements | → | Main part of measurements | ← | Supplementary measurements |
|---|---|---|---|---|
| Water levels<br>Bed levels | | Flow measurements<br>(or discharge measurements) | | Sediment transport |

In this course the following items are discussed:
– water levels (Chapter 2)
– bed levels (Chapter 3)
– discharge measurements (Chapter 4). Special attention is paid to the velocity area method and the determination of the rating curve
– sediment-transport (Chapter 5)
– flow measurement structures (Chapter 6). The structures are classified and the fields of application are mentioned
– hydrological networks (Chapter 7)
– organization of a survey (Chapter 8)

In this course all chapters on measurements are concluded with a section, which gives the relevant International Standards, prepared by ISO/TC 113, Hydrometry. Annex I gives a short description of the International Organization for Standardization (ISO).

*Quick Index of survey methods and instruments*
The most commonly used methods and instruments, used in river surveys and in hydrological networks are listed as follows:

| | |
|---|---|
| water levels | wind effects |
| | staff gauges |
| | float operated gauges |
| | pressure transducers |
| | bubble gauges |
| | ultrasonic sensors |
| | peak level indicators |
| | various recorders |
| | shaft encoders |
| | chimney effect |
| positioning | Global Positioning System, GPS |
| | differential GPS |
| | range finder |
| | sextant |
| bed levels | sounding rod and cable |
| | echo sounder |
| | propagation speed of a sound wave |
| discharges | velocity area method with: |
| | – current meters |
| | – floats |
| | – electro magnetic sensors |
| | wet-line correction |
| | cable way systems |
| | moving boat method |
| | Acoustic Doppler Current Profiler, ADCP |
| | slope area method |
| | Manning's coefficient |
| | dilution method |
| | stage discharge method |
| | acoustic method |
| | electro magnetic coil method |
| | pumping stations |
| sediment transport | various samplers |
| | dune tracking with echo sounders |
| flow measurement structures | weirs, flumes and gates ` |
| | free and submerged flow |
| | modular limit |
| | head-discharge equations |
| | fishways |
| standardization | ISO standards |

All these subjects are discussed in this book.

CHAPTER 2

# Water levels

## 2.1 PURPOSE

Water levels may be considered the basis for any river study. Most kinds of measurements – such as discharges – have to be related to river stages. However, in reality the discharge of a river is a better basic information than the water level, and if it could be possible to measure this discharge daily or even several times a day at many places, this would be preferable.

Water levels are obtained from gauges, either by direct observation or in recorded form. The data can serve several purposes:
– By plotting gauge readings against time over a hydrological year the *hydrograph* for a particular gauging station is obtained. Hydrographs of a series of consecutive years are used to determine *duration curves*, indicating either the probability of occurrence of water levels at the station considered, or, by applying rating curves, indicating the probability of occurrence of discharges.
– Combining gauge readings with discharge values, a stage discharge relation can be determined, resulting in a *rating curve* for the particular station under consideration.
– From the readings of a number of gauges, observed under steady flow conditions and at various stages, *stage relation curves* can be derived.
– Apart from use in hydrological studies and for design purposes, the data can be of direct value for other purposes such as, for instance, navigation, flood prediction, water management and waste water disposal.
Examples of a hydrograph, a rating table and stage relation curves are given in Section 2.8.

## 2.2 THE WATER LEVEL GAUGING STATION

The stage of a stream or lake is the height of the water surface above an established datum plane. The water surface elevation referred to some

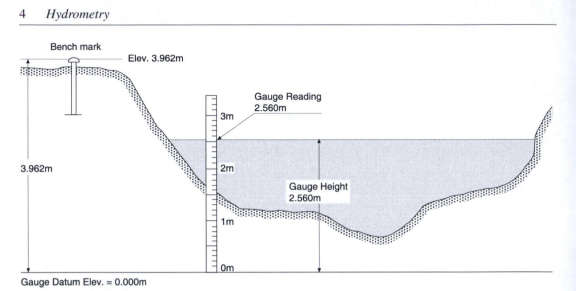

Bench mark

Elev. 3.962m

3.962m

Gauge Reading
2.560m

3m

2m

Gauge Height
2.560m

1m

0m

Gauge Datum Elev. = 0.000m

Figure 2.1. Definition sketch gauge datum and gauge reading (after Water Survey of Canada, 1984).

gauge datum is called the gauge height or stage. Stage or gauge height is usually expressed in metres and centimetres.

A record of stage may be obtained by systematic observations of a staff gauge or with an automatic water level recorder. The advantages of the staff gauge are the low initial cost and the ease of installation. The disadvantages are the need for an observer and the low accuracy. For long term operation the advantages of the recording gauge are far more than those of the staff gauge.

Hence the use of the non-recording gauge as a base gauge is not recommended.

To obtain accurate and reliable stage data, the station gauge and bench-marks must be referred to a fixed datum.

The datum may be a recognized datum plane, such as mean sea level MSL, or an arbitrary datum plane chosen for convenience. An arbitrary datum plane is selected for the convenience of using relatively low numbers for gauge heights, or to eliminate minus values of gauge heights. When a weir or flume is used, the gauge datum is usually set at the elevation of zero flow (crest level).

Each gauging station requires at least two bench-marks or reference marks; that is, permanent points of known elevation, that are independent of the gauge structure. The gauge datum is periodically checked by levelling from the bench-marks to the gauges at the station, at least once a year as a routine.

A universal datum plane does not exist.

Most countries have their own horizontal datum plane. As a consequence, direct comparison of gauged water levels in neighbouring countries is not always possible.

The national datum plane in the Netherlands is called NAP (= Normaal Amsterdam Peil), which is about mean see level in Amsterdam, years ago.

The Belgium datum plane is 2.34 metres lower. The Danish and French datum planes are 0.14 m and 0.13 m higher than NAP, the German Normall Null is 0.02 m higher than NAP. In many countries sloping datums are used for main river systems.

## 2.3 SELECTION OF GAUGE SITES

The network of gauges along a river should be arranged so that water level information at any place along that river can be gathered by means of interpolation of the gauge records. Gauges should be placed where a change occurs in the water level gradient, the discharge, or in general, in the character of the river. At the same time the river sections between two gauges should not be too long. The selection of a gauge site and the installation of a gauge requires thorough knowledge about the hydraulic and morphological phenomena in a river. It also requires knowledge about ship movements, etc. For the gauge site and installation of the gauge, the following requirements should be met:
– the site should be accessible to a gauge reader, who can easily read the gauge, and (unless an automatic recording gauge is used) a gauge reader should be available at all time;
– even during extreme low water levels the gauge should still be in open connection with the river and not be dried;
– even during extreme high water levels the gauge should not be over-flowed;
– damaging of the gauge by ships, floating debris or slidings off the river bank should be prevented;
– the location of the gauge site must be so that no influence is felt of backwater effects due to confluences etc. Preferably the location should be chosen just upstream of a control section so as to avoid the influence of local scour and sedimentation;
– one or two levelled bench-marks should be near, for a regular check of the gauge datum.

The required frequency of gauge readings depends on the fluctuation of the levels. When these fluctuations are small, one reading a day can be sufficient, but if great fluctuations occur, three or more readings a day are required. At places with very rapid changes in water levels, hourly readings should be taken, but it is preferable that continuous readings be taken by an automatic gauge.

Gauges on lakes and reservoirs are normally located near the outlet, but upstream of the zone where an increase in velocity causes a drawdown in water level, which is the case with weirs and flumes.
Gauges in large bodies of water should also be located so, that the effects of strong winds do not generate to much uncertainty.

The change in water level caused by strong winds is given by the following equation:

$$Z = \alpha \cdot v^2 \cdot \ell \cdot \cos \phi / d \tag{2.1}$$

where
$Z$ = difference in water level between two points, $P_1$ and $P_2$ (m)
$\alpha$ = a coefficient ($\sec^2$/m)
$v$ = wind speed (m/s)
$\ell$ = distance between $P_1$ and $P_2$ (m)
$\phi$ = angle between wind direction and the line $P_1 P_2$
$d$ = water depth (m)

The coefficient $\alpha$ depends on conditions of windbreak
– wide canals     $\alpha = 0.20 \cdot 10^{-6} \sec^2$/m
– wide lakes      $\alpha = 0.35 \cdot 10^{-6} \sec^2$/m,  with  8  m/s $< v <$ 18  m/s
(Beaufort 5–7)

The distance along a river between main stations may amount to some tenths of kilometers. Along the River Rhine it varies between 10 and 30 km; on the Mississippi it is about 50 km; along the Niger it varies from 50 to 100 km. Wherever necessary, the information about water levels can be refined by installing secondary gauges between the main stations.

Gauging stations should conform to the requirements of the relevant sections of ISO 4373 which includes recommendations for the design of the reference gauge, recorder, and stilling well.

## 2.4 DIFFERENT TYPES OF GAUGES

### 2.4.1 *Overview of water level gauges*

Most water level gauging stations are equipped with a sensor or gauge and a recorder. In many cases the water level is measured in a stilling well, thus eliminating strong oscillations.

Table 2.1 gives an overview of the gauges that are most commonly used, indicating also the necessity of a stilling well and the way of reading.

### 2.4.2 *The staff gauge*

The *staff gauge* is the simplest type and very popular. It usually consists of a graduated gauge plate, resistant to corrosion: cast aluminium or enamelled steel. This plate is fixed vertically onto a stable structure, such as a pile, a bridge pier or a wall. Sometimes the gauge is placed in an inclined position, for instance upon a sloping river bank. Inclined staff gauges are not exposed to damage by ships or floating material.

Table 2.1. Overview of water level gauges.

| Type of gauge | Stilling well recommended (Section. 2.6) | Way of reading | |
|---|---|---|---|
| | | Not-continuously, visual observation | Continuously, recorder-equipped (see Section 2.5) |
| Staff gauges | – | yes | – |
| Float operated gauges | Indispensable | – | yes |
| Pressure transducers | Preferable | – | yes |
| Bubble gauges | – | – | yes |
| Ultrasonic sensors | – | – | yes |
| Flood crest gauges | – | yes | – |

They are provided with a graduation making allowance for the side slope *m*. In wavy conditions however, accurate reading of the inclined gauge is difficult. Another disadvantage is that adjustment of the gauge is usually not very easy.

The staff gauge is the only gauge which can be read directly, at any time and without preceding manipulations.

Where the range of water levels exceeds the capacity of a single gauge, additional gauges may be installed on the line of the cross section normal to the direction of flow. The scales on such a series of stepped gauges should have adequate overlap.

Figure 2.2 shows a vertical staff gauge fixed onto a pile along a river.
Figure 2.3 gives the so-called E type vertical staff gauge.
Figure 2.4 shows examples of mounting staff gauges.

Figure 2.2. Vertical staff gague.

Figure 2.3. E-type staff gauge.

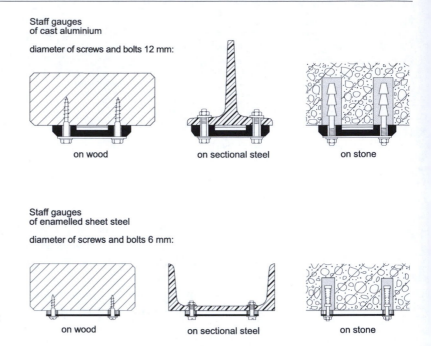

Staff gauges
of cast aluminium

diameter of screws and bolts 12 mm:

on wood          on sectional steel          on stone

Staff gauges
of enamelled sheet steel

diameter of screws and bolts 6 mm:

Figure 2.4. Examples of
mounting staff gauges (after:
Ott, Staffgauges).

on wood          on sectional steel          on stone

### 2.4.3 *The float operated gauge*

The principle of the automatic recording float operated gauge is as follows (Figure 2.5). A float inside a stilling well, which is connected with
the river by an intake pipe, is moved up or down by the water level.
(Fluctuations caused by short waves are almost eliminated.)

The movement of the float is transmitted by the float wheel to a mechanism which records these movements on paper (mechanically), or in a
fixed memory (electronically).

It is strongly recommended to operate the float in a stilling well (Section 2.6).

The functional requirements are:

a) A float operated gauge should permit measurement of stage to be
   made at all levels from the lowest to the highest level expected.
b) Float and counterweight dimensions and the quality of the elements of
   the mechanical device for remote indication should be selected so that
   there is a sufficiently high accuracy.
c) The float should be made of durable corrosion resistant and antifoul-
   ing material. It should be leakproof and function in a truly vertical
   direction. Its density should not change significantly.
d) The float should float properly and the tape or wire should have no
   twists or kinks.

Section 2.6 outlines the function and design of a stilling well.

### 2.4.4 *Pressure transducers*

Pressure transducers are also referred to as pressure sensors, pressure
probes and pressure transmitters.

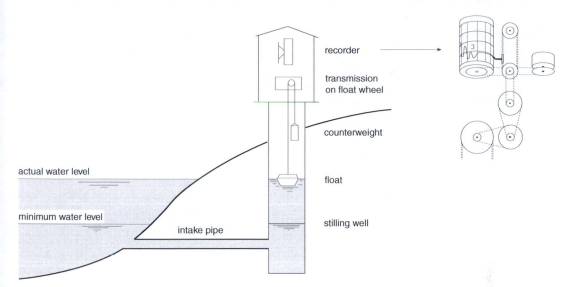

recorder

transmission
on float wheel

counterweight

float

stilling well

actual water level

minimum water level

intake pipe

Figure 2.5. The float
operated recorder.

The water level is measured as a hydrostatic pressure and transformed
into an electrical signal, in most cases with a semi-conductor sensor
(piezo resistive pressure transducer). The measured value corresponds to
the actual water level above the sensor.

In some cases errors are generated due to the varying weight of the
watercolumn (salinity, temperature and sediment content) and fluctuations of the atmospheric pressure.

Pressure transducers are used for the measurement of water levels in
open water, as well as for the continuous recording of groundwater levels.
– *open water*

Pressure transducers are mostly installed in stilling wells. These wells
can be attached to an existing wall, or placed in the embankment in the
same way as for float-operated recorders. The stilling well may not be
closed totally airtight, so as to maintain the atmospheric pressure in it.
– *groundwater*

The transducer is installed in a pipe or a borehole of small diameter and
to large depths (up to 200 m). In this way it is also possible to measure
water levels in not permanently water carrying riverbeds, like wadis.

pressure                    diaphragm

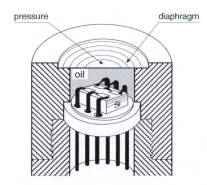

oil

Si

Figure 2.6. Piezo resistive
pressure transducer.

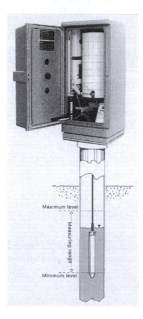

Figure 2.7. Vertical water level gauge (drum recorder) with pressure transducer (after: Seba).

Preferably pressure transducers are compensating for changes in the atmosphere pressure. However, air vented cables (combined with the signal cable) are expensive. In case of submerged self recording systems, there is no cable at all and air pressure needs to be measured separately. Some characteristics of pressure transducers:
– diameter 10 mm < D < 45 mm.
– measuring range: from 0–1.25 m to 0–40 m and more.
– accuracy 0.1% of the full range. It is recommended to check the output of the transducer, using a reference plate connected with the stilling well.
– power supply: Accu 12 V or small batteries.
– output: volts, or milli-amperes (0–20 mA in many cases)
Most pressure transducers are equipped with a data logger, having a storage capacity of at least 10.000 measured values as a standard.

### 2.4.5 *Bubble gauges*

The bubble gauge – also called Pneumatic Water Level Gauge – is a pressure actuated system, based on the measurement of the pressure which is needed to produce bubbles through the bubble orifice against the water pressure. The gauges are primarily used at sites where it would be expensive or difficult to install a float operated recorder or a pressure transducer.

From a pressurized gas cylinder or a small compressor, gas (nitrogen or compressed air) flows over a pressure reducer (for instance a proportioning valve) through the measuring pipe. At the end of this pipe, gas bubbles constantly flow out through the bubble orifice in the water. The pressure in the measuring tube corresponds to the static pressure of the water column above the orifice. This static pressure of the water column is transmitted to a manometer.

Short time variations, like wind waves, should not in any way affect the instrument. Therefore, a damping device is installed at the end of the measuring tube.
Some characteristics of bubble gauges:
– measuring range: from 0–8.00 m to 0–30.00 m
– accuracy: error less than 1 cm over the total range
– power supply: 12V battery (sufficient for 1 year) or connection to 220/110V supply
– output depends on equipment combination
    – use of potentiometer 0–20 mA or volts or a shaft encoder, for connection of data transmission systems
    – connection with an analog strip-chart recorder (several months) or a drum recorder (several days)

### 2.4.6 *Ultrasonic sensor*

Ultrasonic (or acoustic) sensors are used for continuous non-contact level measurements in open channels (see Figure 2.9). The sensor emits

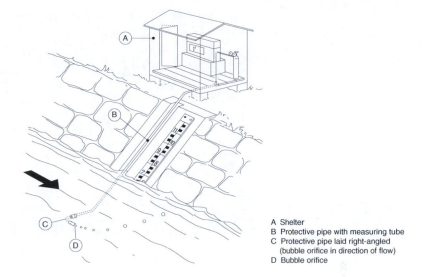

Figure 2.8. Bubble gauge at a riverbank (after: Ott, Pneumatic waterlevel gauge).

A Shelter
B Protective pipe with measuring tube
C Protective pipe laid right-angled
  (bubble orifice in direction of flow)
D Bubble orifice

ultrasonic pulses at a certain frequency. The inaudible sound waves are reflected by the water surface and are received by the sensor. The round-trip time – i.e the time elapsed between transmitting and receiving the echo – is measured electronically and appears as an output signal proportional to level.

In most of the ultrasonic sensors a built-in temperature probe compensates automatically roundtrip time errors which are mainly caused by the temperature coefficient of the speed of sound in air.

Average propagation speeds of sound waves in air and in water:
$C_{air}$ = 330 to 340 m/s
$C_{water}$ = 1450 to 1480 m/s
For open channel flow measurement, the proportional level signal can be modified by a lineariser programmed with the head discharge relation of the particular structure (weir of flume).

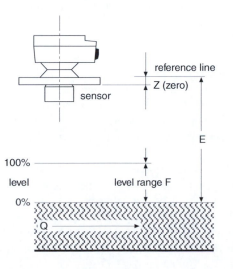

Figure 2.9. Ultrasonic sensor (after: Endress & Hauser, Flowsonic).

Figure 2.10. Ultrasonic sensor
FMU 862 (after: Endress &
Hauser, 1999).

The accuracy of the level measurement depends on the measuring conditions (stable water level) and on the mounting conditions (factor *E/F*). The latter indicates the meter accuracy, which is a function of the *E/F* factor.

### 2.4.7  *Peak level indicators*

After floods a marking line is printed along the banks of rivers and lakes by floating debris, and sediments on the walls of houses, bridges, quay-walls, etc. which have been subject to high water levels.

Apart from a flood survey after an extreme water level, use can be made of special peak level indicators.

Flood crest gauges are very simple instruments to determine top stages during floods at remote or unaccessible locations along rivers. They are read during an inspection after floods. The variety of types depends on local conditions and possibilities and on the inventiveness of the hydrometrist.

Three examples are:

*The Griffin gauge*
The Griffin gauge is a wooden stick painted with soluble paint and fixed in an upright position with the bottom of the stick below water level, preferably in a glass or plastic tube. After the flood, the paint below the highest water level has dissolved, from which the top stage can be deduced.

*Maximum level gauge*
This instrument is a glass-fibre reinforced plastic scale in a plexiglass tube. A transparent self-adhesive red colour tape is fixed to the scale. Water rising in the gauge tube clearly washes off the red colour up to a height reached by the water (the maximum level occured). Replacement of the red colour tape can easily be done by loosening a screw.

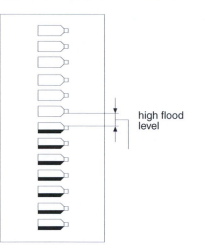

high flood
level

Figure 2.11. Bottle gauge.

*The bottle gauge (see Figure 2.11)*
The bottle gauge consists of a series of bottles placed in a horizontal position and fixed onto a vertical staff. After the flood the maximum water
level is known from the position of the highest bottle containing water.
The bottle gauge should be protected from rain.

The definition of a peak water level in the River Rhine in the Netherlands is as follows:
– the river discharge in Lobith $Q \geq 2500$ m³/s (a discharge which is
   exceeded 25% of the time in an average year)
– the time lapse between two peaks $t > 15$ days

## 2.5  AUTOMATIC RECORDING SYSTEMS

### 2.5.1  *Introduction*

Most types of gauges can be adapted for automatic recording except for
the staff gauge and the flood crest gauges, which require direct observation. Float operated gauges and pressure transducers are usually placed
in stilling wells which dampen fluctuations caused by waves and turbulence, and give protection to the sensor. Normally, a staff gauge is
installed next to a recording gauge to enable comparison of direct readings and the recorded stages.

For recording/registration purposes both mechanical as well as electronic
systems can be used.
   Mechanical recorders are subdivided in analogue recorders and digital
recorders, which can be made to operate unattended for periods from
a couple of weeks to several months. Analogue mechanical recorders
(paper-written drum recorders) give continuous information. Digital
mechanical recorders (for instance the punch type recorders) give information at preselected time intervals.

Table 2.2. Overview of data registration, transportation and transformation.

| Registration method (automatic continuous recorders) | | Transportation | Transformation |
|---|---|---|---|
| Analogue mechanical | on paper (also electrical) | Physical | Digitizing |
| Digital mechanical | on tapes | Physical | Reading and reformatting |
| Electronic loggers | data cartridges data loggers | Physical | Reading and reformatting |
| Telemetry systems | | Transmission line public telephone radio or satellite | Reformatting |

Electronic systems are subdivided in data loggers and telemetry systems. Data loggers are storing the collected data in an electronic memory at the gauging station for a certain period. In telemonitoring systems the collected data are transferred to a central computer, once per day or at real time, by special transmission lines, public telephone line, by radio or satellite.

Table 2.2 gives an overview of data registration, transportation and transformation of collected field data.

The selection of the type of gauge/recorder system is based on both economic and technical considerations. It is obvious that the rather expensive automatic continuous recorders will be chosen only when continuous information is needed, or for remote stations where daily reading is difficult.

The staff gauge is relatively cheap and easy to install, but the facts that reliable observers are required, that salaries must be paid and that human errors may be introduced, must be taken into consideration.

For long term operation, usually automatic continuous recorders will be selected. However, the installation of those systems is only feasible if skilled technicians are available for checking and repairing the recorder at short notice.

### 2.5.2 *Analogue mechanical recorders*

The analogue or autographic mechanical recorder supplies a continuous record on recorder paper of water stage with respect to time. If the recorder is connected to a float-operated gauge, then the movement of the float is converted into a pen movement, which is registered on a paper strip fixed to a rotating drum being moved forward by a clockwork system, thus creating a continuous line on paper. Usually, the gauge height element moves the pen or pencil stylus and the time element moves the chart.

Figure 2.12. Analogue
mechanical recorder.

In the metric system, the range of gauge height scales is from 1:1 to 1:50. Common recording scales are 1:1, 1:2,5, 1:5, 1:10 and 1:20. Time scales vary according to the chart design, but should not be less than 48 mm for 24 hours.

Some analogue mechanical recorders can record an unlimited range in stage by a stylus reversing device or by unlimited rotation of the drum.

Most strip chart recorders (two drums) will be operated for several months without servicing. Single drum recorders (horizontal type and vertical type) require attention at weekly intervals. Figure 2.12 shows a commonly used continuous analogue mechanical recorder. Attachments can be made available to record water temperature or rainfall on the same chart with stage.

Figure 2.13 is a section of a typical strip chart whose gauge scale is 1:5 and whose time scale is 240 mm per day.

It is important that the chart be cut by the manufacturer in such a manner that the direction in which the height is recorded must be in accordance with the machine direction of the paper or, the material with which the chart is made. In the case of strip chart recorders, however, the time scale is always in the machine direction since it is impossible to do otherwise. Paper should be used whose length and width remain relatively unchanged by humidity changes.

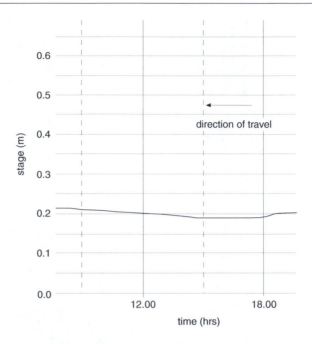

Figure 2.13. Section of typical strip chart analogue recorder.

In addition, the chart division lines must be clearly marked. On the height scale the lines should be numbered at least every metre and on the time scale at least every six hours.

Records from paper-written recorders cannot be read by a computer. Hence the graphs need to be transformed manually, using an electronic digitizer connected to a computer.

Figure 2.14. Digital mechanical recorder.

### 2.5.3 *Digital mechanical recorders*

Examples of digital mechanical recorders are the punch type recorder and magnetic tape recorder. Both types are still used in some locations; however, they are less popular than paper-written recorders and electronic systems. In this section the punch type recorder is discussed.

The punch type recorder is a battery operated, slow speed, paper tape punch which records a four digit number on a 16-channel paper tape at preselected time intervals (see Figure 2.14).

The stage is recorded by the instrument in increments of 0.001 m from zero to 9.999 m and transmitted to the instrument by rotation of the input shaft. Shaft rotation is converted by the instrument into a coded punch tape record that is simple enough to be read directly from the tape.

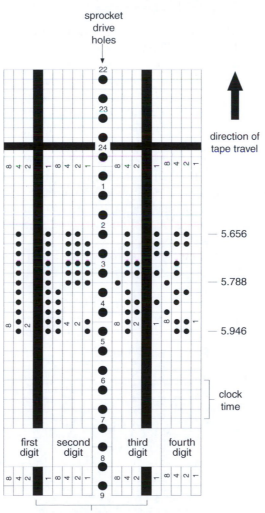

Figure 2.15. Section of a digital tape (after: Water Survey of Canada, 1983).

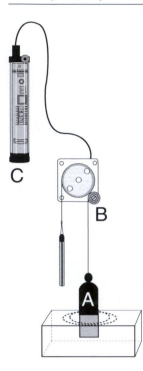

Figure 2.16. A typical shaft
encoder (after: Ott).
a) float,
b) float pulley and disk
   (optical, magnetic or
   otherwise),
c) datalogger.

The code consists of four groups of four punches each. In each group, the first punch represents '1', the second '2', the third '4', and the fourth '8'. Thus a combination of up to three punches in a group represents digits from one to nine, with a blank space for zero, and the four groups of punches represent all numbers from 1 to 9999 (see Figure 2.15).

Mechanically punched tape is the most practical for field use under widely varying conditions of temperature and moisture. Electronic translators are used to transform the 16-channel punch tape records to a tape suitable for input into a computer.

Over the years, punch and magnetic type recorders go out of use.

### 2.5.4 *Electronic recorders*

The data logger is the electronic successor of the paper-written recorder for the measurement and registration of (hydrological) parameters. Measured values – such as a pressure or the position of a float – are converted into electrical signals (analogue or digital). A commonly used industrial standard range is 0-20 mA and 4-20 mA. The measured data are stored in a re-usable solid state memory. The electronic circuits are designed for low power consumption (power supply by batteries). This makes data loggers ideal for application at remote sites.

Modern data loggers have various inputs, and are capable of storing inputs at preset times. Memory cards are available for various storage capacities varying from 10,000 to 50,000 readings.

A read-out of the stored data can be done as follows:
- exchange the full memory card, and read it out by means of a card reader
- transmission of the data by means of a modem and a telephone line to a computer (telemetry)

Figure 2.17. Example of
measured water levels, stored
as ASCII-file (after: Ott).

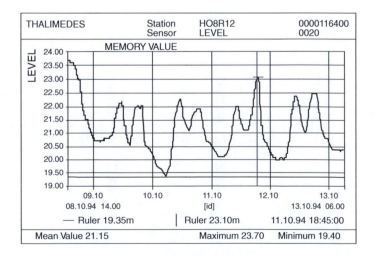

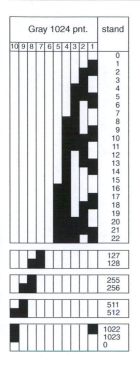

Figure 2.18. The Gray code (1024 numbers) (after: MCB).

Most electronic recorders have a display, showing the read-out of the actual parameters value, date and time, etc.

Electronic data loggers may be used for storing data from various observations such as meteorological measurements, measurements of water levels, discharges and water quality measurements.

Telemetry systems – transmission of data by telephone, radio or satellite – are very popular nowadays.

### 2.5.5 *Shaft Encoders*

A float operated shaft encoder is used for the continuous monitoring and storage of groundwater and surface water levels (see Figure 2.16).

The float pulley is put into motion via the float and the float cable. The shaft encoder is an electromechanical device, converting the rotation of the pulley-shaft in an electrical signal by means of a so-called code disk, which is connected to the pulley.

An optical code disk is composed of the following parts:
– a light source
– transparent disk with an untransparant cartridge (see Figure 2.19)
– a photo-sensitive cell
– electronic network to amplify the electric signal generated by the photo-sensitive cell during exposure

The disk is composed of a number of concentrical loops consisting of transparent and nontransparant blocks. In the centerline of the disk a binary code number can be read, in most cases according to the Gray code (see Figure 2.18).

Figure 2.19. An optical code disk with ten loops (after: MCB).

The electronic level data are transmitted to the data logger in preset intervals.

The shaft encoder's characteristics are:
- appropriate for stand-alone operation
- resolution 1 mm for a range of 20 metres (disk with magnetic strips)
- power supply for a system operation up to 15 months at hourly measuring/storage interval
- simple battery change
- RS 232 interface for data transfer via modem to ASCII-Files

## 2.6 FUNCTION AND DESIGN OF A STILLING WELL FOR FLOAT OPERATED GAUGES

### 2.6.1 *Description of the float system*

The function of a stilling well is to accommodate the water level recorder and protect the float system, to provide within the well an accurate representation of the mean water level in the river and to damp out natural oscillations of the water surface. The function of the intake is to allow water to enter or leave the stilling well so that the water in the well is maintained at the same level as that in the stream under all conditions of flow, and to permit some form of control with which to limit lag and oscillating effects within the well.

Stilling wells are made of concrete, reinforced concrete, concrete block, concrete pipe, steel pipe and occasionally wood. They are usually placed in the bank of the stream (Figure 2.20). Sometimes they are placed directly in the stream and attached to bridge piers or abutments. The stilling well should be long enough for its bottom to be at least 0.3 m below the minimum stage expected and its top sufficiently high to measure the level of the 50-year flood.

The inside of the well should be large enough to enable free operation of all the equipment to be installed (float and counterweight). When placed in the bank of the stream the stilling well should have a sealed bottom so that groundwater cannot seep into it nor stream water leak out.

The stilling well must be installed in a truly vertical position. The free space around the float should be 7.5 centimetres (or more for pipe lengths longer than 6 metres).

Water from the stream enters and leaves the stilling well through the intake so that the water in the well is at the same level as the water in the stream. If the stilling well is in the bank of the stream, the intake consists of a length of pipe connecting the stilling well and the stream.

The intake should be at an elevation of the least 0.15 m lower than the lowest expected stage in the stream, and at least 0.15 m above the bottom of the stilling well to prevent silt built-up from plugging the intake. In cold climates the intake should be below the frostline. If the well is placed in the stream, holes drilled in the stilling well may act as an intake, replacing the intake pipe.

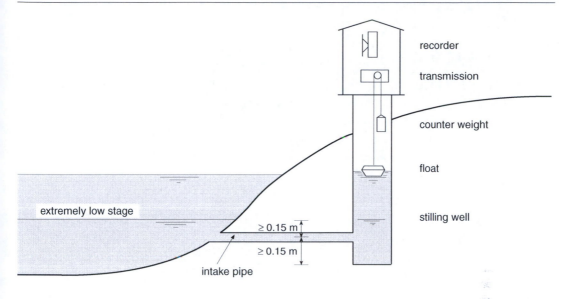

recorder

transmission

counter weight

float

stilling well

extremely low stage

≥ 0.15 m

≥ 0.15 m

intake pipe

Figure 2.20. Principle sketch of a stilling well.

In this section special attention is given to the following items:
– counterweight
– diameter of float, and dimensions stilling well
– diameter intake pipe
– flushing of the intake
– ice protection
– instrument housing
– example of a design
– errors

*Counterweight*
Care should be taken to ensure that if the float is rising, its counterweight does not land on top of the float, but stays well above it or passes the float. If a high degree of accuracy is required, the counterweight should not be allowed to become submerged over a part of the operating range since this will change the submergence rate of the float and thus affect the recorded water level. This systematic error may be prevented by:
– extending the stilling well pipe to such a height that the counterweight neither touches the float wheel at low stage nor the water surface at maximum expected stage, or by
– locating the counterweight inside a separate water tight and water free pipe.

*Diameter of float, and dimensions stilling well (see Figure 2.21)*
Float operated water level recorders need to overcome a certain initial resistance, which is due to friction in the recorder and on the axle. This can be expressed as a resisting torque, $T_r$, on the shaft of the float wheel. If the counterweight exerts a tensile force, $F$, on the float pipe, this force must increase or decrease by $\Delta F$ before the recorder will operate.

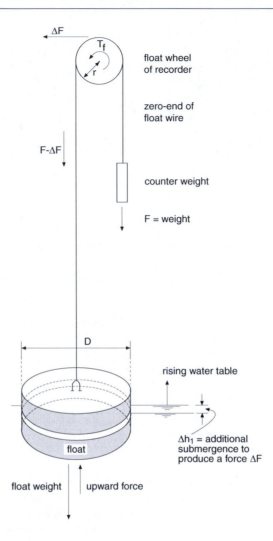

Figure 2.21. Forces acting on a float tape (after: Kraijenhoff van de Leur, 1972).

$$\Delta F > T_f / r \qquad\qquad (2.2)$$

where
$\Delta F$ = change in tensile force on float wire between float and float wheel (N)
$r$    = radius of the float wheel (m)
$T_f$  = resisting torque due to friction on the float wheel axle (Nm)

For a rising water level in the well, a decrease in the tensile force, $\Delta F$, is required, which is possible only if the upward force acting on the submerged part of the float increases. Consequently, the float has to lag behind the rising water level by a distance $\Delta h_1$, so that the volume of the submerged float section will increase by

$$\Delta V = \frac{\pi}{4} D^2 \, \Delta h_1 \qquad\qquad (2.3)$$

where $D$ equals the diameter of the float. According to Archimedes' law, the upward force will increase linearly with the weight of the displaced volume of water, hence

$$\Delta F = \frac{\pi}{4} D^2 \; \Delta h_1 \; \rho g \qquad (2.4)$$

Substitution of Equation (2.4) into Equation (2.2) and expressing the resistance force $T_f/r = \Delta F$ shows that the friction in the recorder and on the axle causes a registration error of the water level, also referred to as a lag.

$$\Delta h_1 = \frac{4 \cdot \Delta F}{\varrho g \, \pi D^2} \qquad (2.5)$$

This float lag causes a systematic error $\Delta h_1$; a rising water level is always registered too low and a falling water level too high.

This systematic error can be limited by making the float diameter $D$ or the radius of the float wheel $r$ sufficiently large. In the following example the float diameter is calculated.

The float diameter follows from:

$$D \geq \sqrt{\frac{4 \cdot \Delta F}{\varrho \cdot g \cdot \pi \cdot \Delta h_1}} \qquad (2.6)$$

Once the float diameter $D$ has been determined, the dimension of the stilling well must be such as to allow unrestricted operation of the equipment installed in it. Clearance between walls and float must be at least 0.075 m. Supposing the counterweight is under all circumstances operating above the float, the stilling well must have a diameter $d_w = D + 0.15$ m.

The level of the stilling well top should fulfil the following two conditions:

1. the recorder, to be placed on the top of the stilling well, should be at least at eye-height, resulting in the recommendation: stilling well top at 1.50 m above field level.
2. the float wheel shaft, located at a vertical distance $e_4$ above the stilling well top, should be in a sufficiently high position:
   - for the maximum water level the counterweight should not be submerged
   - for the minimum water level the counterweight should not contact the float wheel

Designing a stilling well:  start with condition 1
                            check condition 2

*Diameter intake pipe*

The primary purpose of the stilling well is to eliminate or reduce the effects of surging water and wave action in the open channel. Therefore the cross sectional area of the intake should be small. However, the loss of head in the intake pipe (diameter $d_p$) during the estimated maximum rate of change in stage should be limited to a few millimetres. This head loss causes another systematic error $\Delta h_2$: a rising water level is always recorded too low and a falling water level too high. No firm rule can be laid for determining the best size of intake. As a general rule the total cross sectional area of the intake should not be less than 1 percent of the cross sectional area of the well ($d_p \geq 0.1\, d_w$).

The lag for an intake pipe can be estimated for a given rate of change in stage $dh/dt$ assuming that the rate of change in the canal equals that in the stilling well.

The friction losses in a pipe are expressed by the Darcy-Weisbach equation:

$$\Delta h_2 = \lambda\ \frac{L}{d_p}\ \cdot\ \frac{v_p^2}{2g} \tag{2.7}$$

in which
$\Delta h_2$ = friction loss/lag (m)
$\lambda$ = friction coefficient as a function of viscosity and pipe roughness (–)
$L$ = intake length (m)
$d_p$ = diameter intake pipe (m)
$v_p$ = mean velocity in the intake pipe (m/s)
$g$ = acceleration due to gravity, $g = 9.81$ m/s$^2$

When we assume the rate of change $dh/dt$ at both sides of the intake pipe is the same, then:

$$\frac{dh}{dt} = \left(\frac{d_p}{d_w}\right)^2\ \cdot\ v_p, \quad \text{or } v_p = \frac{dh}{dt}\left(\frac{d_w}{d_p}\right)^2 \tag{2.8}$$

in which
$dh/dt$ = rate of change in stage (m/s)
$d_w$ = diameter stilling well

Taking $\lambda = 0.02$ and substitution of Equation (2.8) into Equation (2.7) gives:

$$\Delta h_2 = \frac{0.01L}{g}\ \cdot\ \frac{(d_w)^4}{(d_p)^5}\ \cdot\ \left(\frac{dh}{dt}\right)^2 \tag{2.9}$$

The intake pipe diameter follows from:

$$d_p = \sqrt[5]{\frac{0.01\, L\ \cdot\ d_w^4\ \cdot\ \left(\dfrac{dh}{dt}\right)^2}{g\ \cdot\ \Delta h_2}} \tag{2.10}$$

Where practical, the intake pipe should be laid level and straight on a suitable foundation which will not subside, or at a constant gradient with the highest point at the stilling well. The pipe should enter the stream perpendicular to the direction of flow, and where it terminates in a concrete wall it should be set flush with the wall.

In case the intake pipe does not terminate, flush with the side slope while protruding a certain length into the canal, then the measured water level will be subject to drawdown as a result of the 'chimney-effect' (high velocity over the end of the pipe). This drawdown can be written as:

$$\Delta h_3 = \alpha_d \cdot \frac{v^2}{2g} \tag{2.11}$$

where
$\Delta h_3$  =  drawdown of chimney effect
$\alpha_d$  =  coefficient $1.0 < \alpha_d < 1.3$
$v$  =  flow velocity at the end of the intake pipe
   The drawdown can be prevented by fitting the end of the intake pipe with a static tube (principle of a Pitot-tube). See also section 2.6.4.

*Flushing of the intake*
In most stilling wells, the intake pipes will require periodical cleaning, especially those in rivers carrying sediments. Permanent installations can be equipped with a flushing tank as shown in Figure 2.22. The tank is filled either by a manual pump or with a bucket, and a sudden release valve will flush water through the intake pipe, thus removing the sediment. For tightly clogged pipes and on temporary or semi-permanent structures, a sewer rod or 'snake' will usually provide a satisfactory way of cleaning.

*Ice protection*
In cold climates the well should be protected from the formation of ice which could restrict or prevent the free operation of the float operated system. The usual means of preventing the formation of ice in the well during cold weather are: subfloors, heaters or oil.

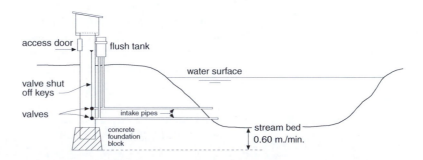

Figure 2.22. Example of an intake pipe system with flush tank (after: Bos, 1989)

Subfloors are effective if the station is placed in the bank and has plenty of fill around it. If the subfloor is built in the well below the frost-line in the ground, ice will not normally form in the well as long as the stage remains below the subfloor. Holes are cut in the subfloor for the float tape and counterweight to pass through. Subfloors prevent air circulation in the well and the accompanying heat loss.

An electric heater or heating light with reflectors may be used to keep the well free of ice. The cost of operation and the availability of electric service at the gauging station are governing factors. Heating cables are often placed in intake pipes to prevent the forming of ice. When oil is used the oil surface will stand higher than the water level in the stream. When oil is added, then we need to reset the recorder (A problem arises if oil is gradually lost).

*Instrument housing*

Some form of housing is essential to protect the recording and measuring equipment from the elements or from unauthorized attention or vandalism, and also to give shelter to the servicing and gauging staff who operate the station. The size and quality of the housing adopted will depend on conditions at the site and on the type and range of equipment to be installed in it, but in general the minimum required is a well-ventilated, weatherproof and lockfast hut set on a stable foundation and of such dimensions as will permit normal servicing of the recorders.

### 2.6.2 *Example of the design of a float operated system*

The water level in an irrigation main canal is to be measured with a float operated recorder. The stilling well is to be placed in the embankment.

a) *Field information*
  – bottom level canal MSL + 4.50 m
  – field level MSL + 8.50 m
  – canal side slope $m = 1.5$
  – maximum water level MSL + 7.80 m
  – minimum water level MSL + 5.90 m
  – maximum level rise or fall $dh/dt = 0.001$ m/s
  – information on the internal friction recorder
    – resisting torque axle float wheel $T_f = 0.015$ Nm
    – float wheel diameter $d_{fw} = 0.116$ m
  – dimensions float and counter weight
    $e_1 = 0.05$ m and $e_2 = 0.15$ m (Figure 2.23)
  – intake pipe not flush, (see chimney effect, Figure 2.26)
  – float wheel at $e_4 = 0.15$ m above stilling well top

b) *Design stilling well, step by step*
  – float diameter $D$ (maximum allowable lag $\Delta h_1 = 0.0015$ m).

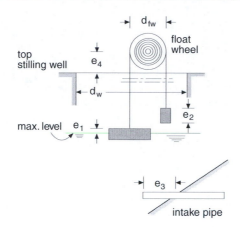

Figure 2.23. Details design
float operated recorder.

$$D = \sqrt{\frac{4\ \Delta F}{\varrho \cdot g \cdot \pi \cdot \Delta h_1}}$$

- stilling well diameter $d_w$
- level and length $L$ intake pipe (minimum water level − 0.15 m)
- level of stilling well bottom (minimum water level − 0.30 m)
- intake pipe diameter $d_p$ (maximum lag $\Delta h_2 = 0.0015$ m).

$$d_p = \sqrt[5]{\frac{0.01\ L \cdot d_w^4 \cdot \left(\dfrac{dh}{dt}\right)^2}{g \cdot \Delta h_2}}$$

- level stilling well top and cable length $\ell_c$.

Make a design sketch, giving all dimensions.

c) *Design calculation*
- compute float diameter $D$ from Equation (2.6) resisting force float wheel $\Delta F = T_i/0.5\ d_{fw} = 0.015/0.5 \cdot 0.116 = 0.259$ N

$$D = \sqrt{\frac{4 * 0.259}{1000 * 9.81 * \pi * 0.0015}} = 0.150 \text{ m}$$

- stilling well diameter $d_w = 0.150 + 0.075 + 0.075 = 0.300$ m
- level intake pipe MSL + 5.90 − 0.15 = MSL + 5.75 m
- level stilling well bottom MSL + 5.90 − 0.30 = MSL + 5.60 m
- length intake pipe $L = 0.10 + 1.5 (8.50 − 5.75) + 1.00 − 0.15 = 5.08$ m (Fig. 2.25)
- use equation 2.10 to find the intake pipe diameter

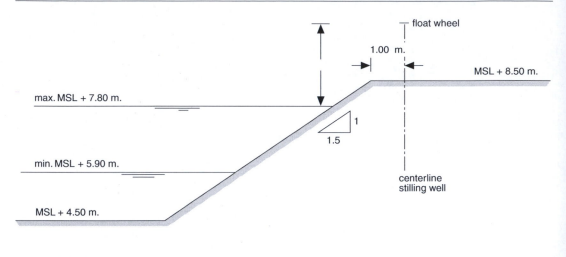

Figure 2.24. Field information design float operated gauge.

$$d_{\mathrm{p}} = \sqrt[5]{\frac{0.01 * 5.08 * (0.30)^4 * (0.001)^2}{9.81 * 0.0015}} = 0.031 \text{ m}$$

– compute level stilling well top and cable length $\ell_c$ as follows: First condition: recorder at eye height → stilling well top 1.50 m above field level. Stilling well top MSL + 8.50 + 1.50 = MSL + 10.00 m. As a consequence: level float wheel MSL + 10.00 + 0.15 = MSL + 10.15 m.
Second condition: free movement counterweight for all water levels, which also depends on the total effective cable length $\ell_c$
  – maximum water level: counterweight not submerged
    $\ell_c \leq [10.15 - (7.80 - 0.05) + \pi \cdot 0.116/2] + [10.15 - (7.80 - 0.15)]$ $\ell_c \leq 5.08$ m
  – minimum water level: counterweight not in contact with float wheel
    $\ell_c \geq [10.15 - (5.90 - 0.05)] + (\pi \cdot 0.116/2) + (0.116/2)$ $\ell_c \geq 4.54$ m
    Second condition is fulfilled for $4.54 < \ell_c < 5.08$ m
    Take $\ell_c = (4.54 + 5.08)/2 = 4.81$ ml → margin in both directions $s = 0.27$ m
    (The higher the stilling well top and the float wheel position, the more is the margin $s$. In this example becomes the minimum level stilling well top MSL + 9.73 m)

### 2.6.3 *Errors with float-operated systems*

The principal sources of error, inherent in a float-operated instrument are float lag, line shift, and submergence of the counterweight. With regard to the algebraic sign of the errors discussed below, a positive (+) sign indicates that the instrument shows a stage higher than the true stage, and a negative (−) sign indicates that the instrument has been under-registering.

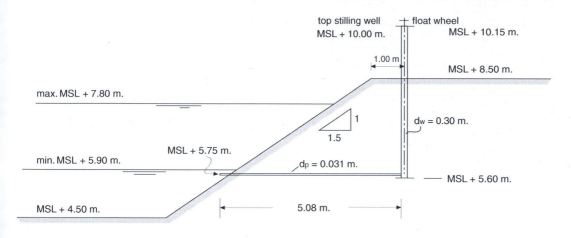

Figure 2.25. Completed design float-operated gauge.

*Float lag (the error $\Delta h_1$)*

If the float operated recorder is set to the true water level while the water level is rising, it will then show the correct water level, as far as float lag is concerned, for all rising stages. For falling stages, however, the recorded stage will be above the true water level (positive error) by the amount of float lag or change in flotation depth of the float. A reverse effect occurs if the original gauge setting is made when the water level is falling. Float lag varies directly with the force ($\Delta F$) required to move the mechanism of the recorder and inversely with the square of the float diameter $D$. The equation for maximum float lag is given in Equation (2.5):

$$\Delta h_1 = \frac{4 \cdot \Delta F}{\varrho g \, \pi D^2}$$

*Example:*

$$\left.\begin{array}{l} \text{Resisting force } \Delta F = 0.1 \text{ N} \\ \text{float diameter } D = 0.08 \text{ m} \end{array}\right\} \Delta h_1 = \pm 0.002 \, \text{m}$$

If the recorder was set at a stationary stage, then the error will be $\Delta h_1 = +0.002$ m for falling stages and $\Delta h_1 = -0.002$ m for rising stages.

*Line shift*

With every change of stage a part of the float tape passes from one side of the float pulley to the other, and the change in weight changes the depth of flotation of the float. The error depends on the magnitude of change in stage $\ell$ since the last correct setting of the recorder, the unit weight $u$ of the tape, and the float diameter ($D$). The error will be positive (+) for a rising stage and negative (−) for a falling stage. The equation for line shift error:

$$\Delta h_4 = \frac{4 \cdot u \cdot \ell}{\varrho g \cdot \pi D^2}$$  (2.12)

*Example*
unit weight of tape $u = 0.002$ kg/m
$\ell = 5.00$ m
$D = 0.08$ m
The line shift follows from Equation (2.12), $\Delta h_4 = 0.002$ m.

*Submergence of the counterweight*
When the counterweight and any part of the float line becomes submerged as the stage rises, the pull on the float is reduced and its depth of flotation is increased. The converse is true when the submerged counterweight emerges from the water on a falling stage.

The errors by float lag and line shift can be reduced by increasing the float diameter $D$. The errors by submergence of the counterweight should be prevented by making the top of the stilling well sufficiently high, or by locating the counterweight in a separate watertight pipe.

### 2.6.4  *Chimney effect with a protruding intake pipe*

The intake pipe shall terminate flush with the side slope of the approach channel and at right angles thereto. The side slope shall be plane and smooth within a distance of ten times the diameter of the pipe (ISO 4359).

In practice the approach channel boundary is not in all cases plane and smooth at the location of the stilling well and intake pipe. Many engineers decide then to enlarge the intake pipe, protruding over a length $L$ in the channel, as shown in Figure 2.26.

As a consequence the pipe operates as a chimney shaft: at the end of the pipe there is an under pressure, resulting in a lowered water level in the stilling well.

Figure 2.26. Stilling well with a protruding intake pipe.

This chimney effect $\delta_h$ is a systematic error: it is a systematic underestimation of the measured water level

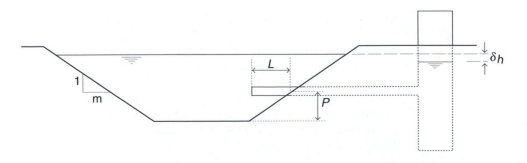

The chimney effect $\delta_h$ is a function of the following parameters
- the flow velocity $v$ at the end of the intake pipe ⎫
- the distance $L$, the length of protrusion           ⎬ strongly
- the distance $P$        ⎱ slightly in a less
- the pipe diameter $D$   ⎰ degree

Experiments in the Hydraulics Laboratory (Dommerholt, 1992) showed maximum values $\delta_h = 1.3 \; v^2/2 \; g$

*Conclusion*
The chimney effect in head measurements using a stilling well with a protruding intake pipe leads to a substantial systematic under estimation of the measured water level.

## 2.7 ACCURACY OF WATER LEVEL MEASUREMENTS

The determination of water levels is subject to errors.
Depending on their nature, errors in measured water levels can be classified in three groups:
1. Errors in the *number of measured values* which may be caused by:
   - failures of the timing mechanism. In the case of analogue mechanical recorders the clock element rotates the chart paper. In the case of the digital mechanical recorder it programs the interval at which the paper tape is punched.
     The accuracy of the clock should be within ±30 seconds per day.
   - incorrect setting of pen or pencil and changes in temperature or humidity can result in an interrupted recording.
     Missing data for a particular measuring station can generally be restored, using stage relation curves (see figure 2.30).
2. Errors in the *magnitude of measured values*
   The nature of these errors can be systematic (e.g., float lag, line shift and chimney effects) as well as random. Staff gauges can be read with an accuracy between 1 and 3 cm depending on the location and wind conditions. For flood crest gauges an accuracy of 5 to 10 cm can be obtained. However ultrasonic sensors may have accuracies of 1 to 2 mm.
   The required accuracy depends on the aim of the measurement. For instance, an error of 1 to 2 cm is acceptable for the determination of stage hydrographs, duration and stage relation curves and for the establishment of rating curves. For top stages and for information relevant to navigation, errors of 5 to 10 cm are acceptable. If readings are taken for the precise determination of local slope of the water surface, however, an accuracy of 0.1 to 0.5 cm may be required, depending on the distance between the gauges. Water levels measured upstream of discharge measurement structures (weirs) require an accuracy of 3 to 4 mm.

Table 2.3. Random errors for different water level gauges.

| Type of sensor | Range of random error |
| --- | --- |
| Staff gauge | 0.010 – 0.030 m |
| Float-operated gauge | 0.002 – 0.004 m |
| Pressure transducer | |
| – low-cost types | 0.010 – 0.050 m* |
| – expensive types | 0.002 – 0.010 m* |
| Bubble gauge | 0.005 – 0.015 m* |
| Ultrasonic sensor | 0.002 – 0.010 m* |
| Peak level indicators | 0.050 – 0.100 m |

* depending on total range of measurement.

3. Errors in zero-setting: determination of the gauge zero by leveling from Benchmarks. It is recommended to verify the gauge zero of the water-level measuring device once per year.
4. Errors in administration are caused by careless filling out of date and time of observation and by exchange of different tapes or papers.

## 2.8  PRESENTATION OF RESULTS OF WATER LEVEL MEASUREMENTS

The required accuracy of water level data depends on the purpose for which the data will be used:
–   Stage records taken along rivers and used for hydrological studies, for design of irrigation works or for flood protection require an accuracy of 2 to 5 cm;
–   Gauge readings taken upstream of flow measurement weirs, used to calculate discharges from the measured heads require an accuracy of 2 to 5 mm. Head discharge relations of weirs are discussed in Chapter 6.

Water levels recorded along rivers are usually sent to a central office, where they are checked for errors and properly filed for future use. Since the development of computers, the processing methods have changed.

Hydrographs, rating tables and stage relation curves are typical presentations of water level data.

*Hydrograph*
A hydrograph is obtained when the stage records of a particular station, or the measured discharges in this location, are plotted against time.

The hydrograph usually covers the hydrological year and also frequently the calender year.

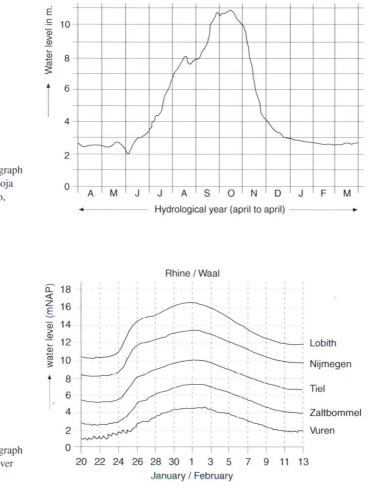

Figure 2.27. Stage hydrograph of the Niger River at Lokoja 1955-1956 (after: Nedeco, 1959).

Figure 2.28. Stage hydrograph of the Rhine and Waal River 1995.

Figure 2.27 is an example of a stage hydrograph of the Niger River, covering a hydrological year that runs from April through March.

Figure 2.28 is a stage hydrograph, covering a two-weeks period during the 1995 floods in the Rhine and Waal River in the Netherlands. The graph shows various hydrographs. The maximum discharge in Lobith was $Q = 12.060$ m³/s (once per 80 years).

*Rating table*
At many gauging stations, discharges are measured during the hydrological year with a fixed frequency of e.g. once or twice per month together with the water level reading.

These data collected over a series of years allow the composition of a rating curve or a rating table for a certain location along the river.

Figure 2.29 is an example of a rating curve for the Ganges River. Combining the data of a rating table and a stage hydrograph, a discharge hydrograph can be made.

Figure 2.30 gives a discharge hydrograph of the Niger River for two consecutive years.

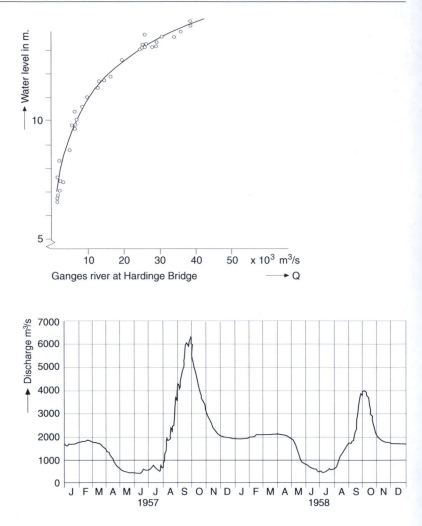

Figure 2.29. Rating table
Ganges River (after: Jansen,
1979).

Figure 2.30. Discharge
hydrograph of the Niger River
(after: Jansen, 1979).

*Stage relation curves*

From the hydrographs of two or more gauging stations along the river,
relationships can be formulated between the steady flow stages at these
stations. These *stage relation curves* are of special interest for rivers in
which the water level fluctuates rather slowly and gradually. They are
frequently used for investigations and calculations. For instance, from
the relationship between the stages at two gauges (of which the zeros
are related to each other or to a reference level) and using the distance
between the gauges, the mean slope of the water surface can be found
at each stage. This slope, in turn, can be used to determine the mean
hydraulic roughness of the river over the reach considered.

Under non steady conditions, the relationship between the stages at
various gauges will be disturbed. In order to arrive at proper relationships
valid for steady conditions, this effect has to be eliminated by omitting
periods of considerable rise and fall from the hydrographs.

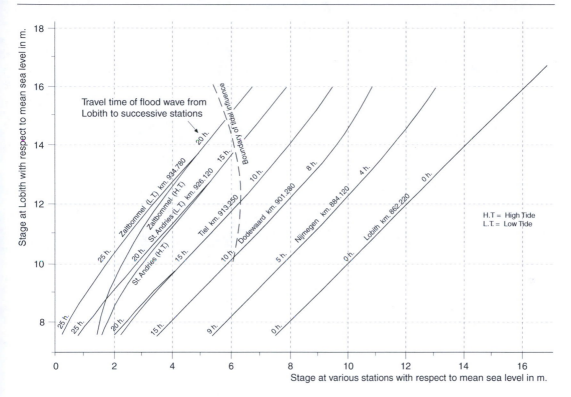

Figure 2.31. Stage relation
curves along the River Waal
(after: Jansen, 1979).

The approximate travel time of flood waves at various stages can be
indicated in the stage relation curves.

Figure 2.31 is an example of a stage relation curve for various gauging stations along the Rhine-Waal River, also indicating the travel times
of flood waves. The distance Lobith-Nijmegen is 22 km, the distance
Lobith-Tiel is 52 km.

## 2.9  INTERNATIONAL STANDARDS

The following ISO standards, dealing with water level measurements and
including groundwater levels in wells and boreholes, are available:

ISO 1100-1      Establishment and operation of a gauging station;
ISO 4373        Water level measuring devices;
ISO 11330       Determination of volume of water and water level in
                lakes and reservoirs.
ISO/TR 14685    Geophysical logging of boreholes for hydrogeological
                purposes.
ISO 14686       Pumping tests for water wells.
ISO 21413       Manual methods for the measurement of a groundwater
                level in a well.

CHAPTER 3

# Measurement of bed levels

## 3.1 INTRODUCTION

The topography of a river bed can be determined by sounding from the water surface, i.e. by measuring the depth of water at a sufficient number of points.

If the sounding operations are intended to form part of a river survey, determination of the exact location $X$, $Y$ of each sounding is essential and sounding levels must be correlated with one another or with a known datum level $Z$.

On a routine basis the measurement of bed levels will take place periodically, for instance once every year. In the case of special investigations the spacing, the timing and the frequency of the measurements will be adapted to their specific aim.

*Sounding procedure*

A complete sounding procedure includes:
–   Preparatory work, which consists of establishing, checking and maintaining the geometric base that enables proper determination of the location of the soundings.
–   The sounding operation itself, consisting of the determination of the location of the sounded points, of the local water level during the sounding and the actual sounding.
–   Processing of the data.

Neglecting or omitting any of the above items, would endanger the success of the whole operation. A decision also has to be made on the density of the network of points to be sounded before the operation takes place.

Obviously, the density is determined by the extent of information required.

The sounding measurements are also referred to as bathymetry (a bathometer measures depths).

The bathymetric survey is an essential part of every field study as it gives information about the bottom configuration, the cross sectional profiles in rivers, and an insight into sedimentation or degradation; it thus gives the basic information to the engineer of the area under study.

The *bathymetric survey* may serve the following purposes:
- navigation: sounding charts are indispensable during low river discharges in non-tidal rivers, and during low tide in tidal rivers;
- discharge calculation: measurements of the cross-sectional profile and velocity measurements give information on the discharges;
- dredging: by sounding the bed before and after dredging, the desired bottom configuration is checked and the volumes can be compared;
- monitoring of morphological developments;
- design of hydraulic structures: design of structures as weirs and bridges need information on the bed level as a function of time.

The bathymetric survey consists of the following components:
- position fixing → what is the location?
- sounding → what is the depth?
- water level → what is the water level during the sounding?

Depending on the circumstances, soundings can be executed either along straight lines in a well determined grid of cross sections or in a free sounding system.

a) *Soundings along straight lines in well determined cross sections*
*Advantages*: the sounded points of the river bed are at approximately fixed locations, thus permitting detailed comparison of soundings performed on various dates; the determination of the location of the sounded points is simple.

*Disadvantages*: the work involved in the geometric base is extensive and needs much care; the geometric base is vulnerable due to the possibility of marks being lost by erosion, by overgrowing vegetation or by sediment deposition. A sounded cross section is an essential element for the determination of river discharge.

*Application*: In rivers, the bathymetric survey consists mainly of cross sectional soundings and soundings along the talway. Soundings can be taken in combination with velocity measurements in order to determine the river discharge or the total sediment transport.

b) *Free sounding system*
*Advantages*: the number of base points needed is restricted and they can be erected at suitable non-vulnerable locations; existing characteristic points may be used as base points.

*Disadvantages*: fast determination of the location of the sounded points is only possible if expensive electronic positioning systems are used; where the position is derived from measurements using a sextant or range finder, the determination of the location takes rather a long time.

Selection of either system should be based on the purpose of the sounding and on the consideration of local conditions such as, for instance, the width of the river, accessibility of the banks, feasibility of establishing and maintaining a geometric base, availability of skilled technicians for servicing electronic equipment, etc.

*Application*: In estuaries and along open coasts the bathymetric survey is carried out along tracks as much as possible perpendicular to the depth contour lines.

## 3.2 POSITION FIXING

a) *Sounding along straight lines*
When the sounding takes place in preselected cross sections, the location of the sounded points is defined by the line of the cross section and the distance between the points and a reference mark. The line of the cross section is indicated on one or both banks by marks, one of which functions as the reference mark. Its location should be known exactly with respect to the geometric base.

In order to keep the survey boat in line, basically two methods can be applied: along a rope, or by sighting.

– Along a rope
  A simple method, which can be applied in rivers with a width not exceeding 150 m, is to stretch a steel wire rope across the river using a winch, if necessary. In fact, the maximum river width for using cables, depends on the height of the banks and on security/safety measures. Tags or tape marks are attached to the rope to indicate the places where the soundings are to be made. Obviously, this method can only be used when there is minimum traffic in the river.

– By sighting
  If the method 'along a rope' cannot be applied, the survey boat has to be kept in line by sighting. For that purpose two marks are needed on the same bank, indicating the line, to enable sighting by the surveyors working on the boat. During measurements the marks can be made visible from the river by using beacons or flags.

The measurement of the distance of each sounding point to the reference mark should preferably be made from the boat directly, with a tape, by optical instruments or by electronic instruments (*direct method*). Indirect methods can be applied if the distance of the sounded points to the reference mark cannot be measured directly, for instance if pegging out of a line of sufficient length is impossible due to the local conditions on the river banks. An example of direct and indirect determination of the location is shown in Figure 3.1, where the angle $\alpha$ is measured by a sextant from the point $A$ to a base $BC$ defined by two reference marks on the bank.

b) *Sounding in a free sounding system*
Applying the free system, a number of visible base points are needed, the locations of which are known exactly. Suitable characteristic points such as towers can be chosen. The location of each point, is, determined by the simultaneous measurement of two angles (by two surveyors), as is shown in Figure 3.2.

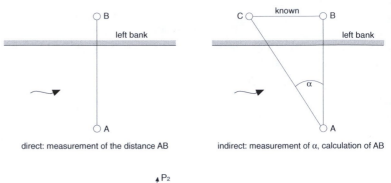

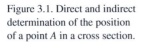

Figure 3.1. Direct and indirect determination of the position of a point *A* in a cross section.

direct: measurement of the distance AB

indirect: measurement of α, calculation of AB

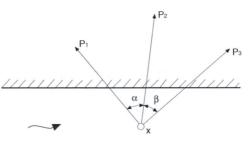

Figure 3.2. Method of resection.

To draw the point *X* on the sounding chart, a station pointer can be used. It consists of three rulers radiating in variable directions from one centre. After setting the appropriate angles α and β between the rulers, the instrument can be shifted over the chart until the three base points *A*, *B* and *C* coincide with the three rulers; the position of *X* is then marked in the centre.

To simplify the plotting of many points during or after observations, a chart of arcs can be constructed, each arc representing the locus of a point making a certain angle between two of the base points (*A* and *B*, or *B* and *C*). The intersection of both loci gives the position of the boat, as is shown in Figure 3.3.

## 3.3  GLOBAL POSITIONING SYSTEM, GPS

In this section use has been made of two brochures written by Jeff Hurn for Trimble Navigation, the Lecture Notes GIS by Loedeman (et al) and a case study reported by Loedeman (see References).
GPS is divided into:
– the spatial segment, formed by 24 satellites
– the user segment, receivers with GPS antennas

Global Positioning System (GPS) was developed by the U.S. Department of Defense (DoD) as a worldwide positioning resource for both military and civilian use. GPS is based on a worldwide three dimensional reference system and the constellation of twenty four satellites orbiting the

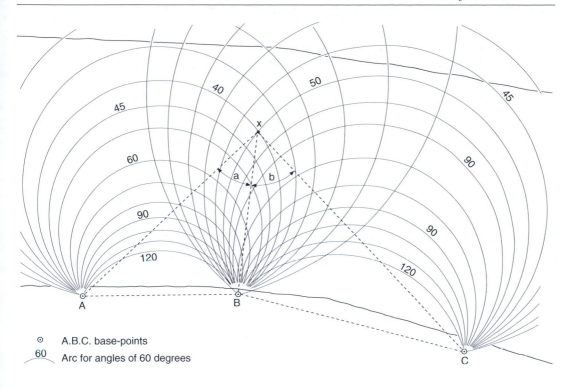

○     A.B.C. base-points

60    Arc for angles of 60 degrees

Figure 3.3. Chart of arcs on three base points (after: Jansen, 1979).

earth over twenty thousand kilometres up. The satellites act as reference points from which receivers on the ground triangulate their position. The satellites orbits are very accurately monitored by groundstations, operated by US DoD.

In hydrometry GPS is used for several purposes such as the measurement of riverbeds (river alignment and river cross-sections) and for position fixing in a number of discharge measurement methods (conventional velocity-area methods and the Acoustic Doppler Current Profiler, ADCP).

The positional accuracy obtainable ranges from about 100 meter for a low-cost single receiver, to better than 1 cm for a set of two sophisticated receivers operating in differential mode which is more expensive.

The basic concepts of GPS are described by Hurn in a *number of steps*:
1. *Triangulation from satellite positions is the basis of the system*
   – position is calculated in XYZ coordinates from distance measurements to satellites
   – mathematically we need three satellites to determine the exact position:
   – one measurement tells us we must be somewhere on an imaginary sphere that is centred on the satellite and that has a known radius.
   – a second measurement narrows down our position to the intersection of two spheres, which is a circle.

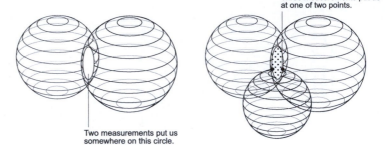

Three measurements put us
at one of two points.

Two measurements put us
somewhere on this circle.

Figure 3.4. Range to three
satellites (after: Hurn 1993).

- a third measurement puts us to the intersection of two circles, which are two points (see Figure 3.4). One of the two points is rejected by GPS as it is not realistic. The remaining point gives the correct position. Conclusion: three measurements are enough if we reject ridiculous answers.
- measurement of a fourth satellite is required for another reason: to cancel out receiver clock errors.

2. *Measuring the distance from a satellite*
   - the GPS system works by timing how long it takes a radio signal to reach the receiver from a satellite and then calculating the distance from the measured speed and the travel time: speed of light x time = distance.
   Radiowaves travel at the speed of light: about 300.000 km per second, so that it takes about 0.06 seconds for the radio message to reach us. Most receivers can measure time with an error of $1.10^{-9}$ second.
   - by synchronizing the satellites and receivers, they are generating the same message (code) at exactly the same time. We know how long it takes for the satellite's signal to get to us by comparing both clock-codes.

3. *Getting perfect timing*
   - satellites have four clocks which are unbelievably precise and extremely expensive. Receivers have moderately accurate clocks. As a consequence the synchronisation will be imperfect.
   - an extra measurement (the fourth satellite) will cancel out any error, applying the well known 'four equations and four unknowns' problem (three ranges and one clock offset).

4. *Knowing where a satellite is in space*
   - GPS satellites are so high up, their orbits are very predictable, even in advance. Moderate ground receivers are equipped with a computer memory, which tells them where in the sky each satellite will be at any given moment.
   - minor variations in orbits are measured constantly by US DoD and that data is transmitted from the satellites themselves.

5. *Ionospheric-atmospheric delays and other error sources*
   - as the GPS signal travels through the earth's ionosphere and atmosphere (variations in weather conditions) it gets delayed, causing a range error.

- some of these errors can be eliminated with mathematics and modelling.
- most errors will be cancelled out by operating two receivers in concert in stead of one autonomous receiver. This approach is called Differential GPS.

*Differential GPS involves cooperation between two receivers.* The GPS differential positioning, DGPS, is a technique in which two receivers are used: one installed in a reference station of known position, and one which occupies the new points to be determined. If two receivers are fairly close together, the signals that reach both of them will have travelled virtually the same slice of atmosphere and will have virtually the same delays. Therefore most of the errors will be common to both receivers, and we can have one receiver (the stationary) to measure those errors and provide that error information to the other receiver (the wandering one). All we have to do is put the reference receiver on a point that has been very accurately surveyed (a known position $X$, $Y$, $Z$).

The typical errors using Differential GPS are far smaller, or eliminated, compared with Standard GPS. Moreover the latest high-precision receiver can produce still better accuracies.

Differential GPS is used by Coast Guards, in aviation, for natural resources management, etc.

Landsurveyers and hydrometrists are using DGPS to do extremely precise surveys, fixing relative positions with accuracies on the order of centimetres. They do use multiple receivers like the DGPS, discussed here, but with a special GPS technique (see literature, Hurn 1989 and 1993).

In hydrometric surveys the setup of a baseline network is essential. Kinematic DGPS measurements need to be started on a point where the position is known. After a starting procedure, one receiver serves as a static reference station, whereas the other is moved from one observation point to another. At each observation point the antenna has to be static for about one minute. With the latest type of receivers this static period can be reduced drastically.

The mobile receiver has to lock permanently on at least four satellites. The number of satellites that can be received during displacement will vary according the terrain relief. As soon as this number drops below four (e.g. in built up areas or dense forests), a loss of lock occurs and the moving receiver will warn with a signal. In that case the survey has to be restarted at a previously registered observation point (Loedeman, 1993).

*Converting GPS positions into a local reference system XYZ*
The final results of position fixing in hydrometric surveys will be given in a local (mostly national) reference system. Therefore GPS positions must be converted into local $XY$ and $H$ (or $Z$). Heights are always expressed

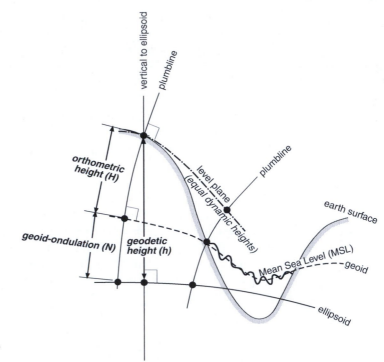

Figure 3.5. Various reference planes (after: Loedeman et al., April 1998).

in reference of an ellipsoid, a geoid or a level plane. Figure 3.5 gives an overview of these reference planes.

From Figure 3.5 we may learn the following aspects:

1. The geoid reference plane is a good approximation of Mean Sea Level (MSL) worldwide. However this plane is rather irregularly shaped at worldscale, and as a consequence it is very complicated to describe the geoid reference plane in a mathematical model.

   For the orthometric height ($H$) the geoid serves as a reference plane.

2. Ellipsoids (slightly flattened spheres) are very good alternatives for the geoid, and rather easy to describe. The best approximation of MSL worldwide is an ellipsoid with 1/298 flattening.

   The geodetic height h refers to the ellipsoid as a reference plane. An overall coincidence of the ellipsoid and the geoid does not exist. The difference between both is defined as geoid undulation, $N = h - H$

   – ellipsoids at a world scale will present large values for $N$ (about 20 metres as a maximum). GPS positions are given in coordinates with reference to the WGS 84-ellipsoid (World Geodetic System, introduced in 1984 by US DoD).

   – ellipsoids at a local scale are the best approximations of the local MSL. In the Netherlands the Bessel-ellipsoid is used. Local ellipsoids cannot be used outside their own specific area. Therefore a variety of many local ellipsoids is needed to describe the earth's shape.

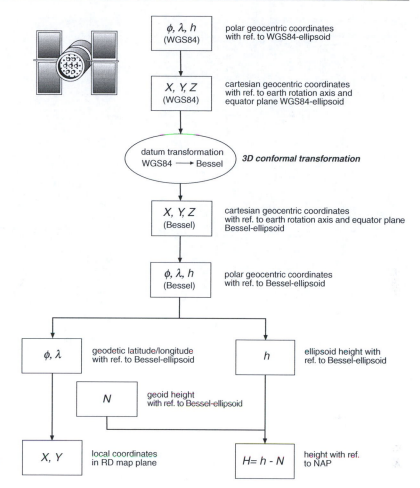

Figure 3.6. Process scheme, converting GPS positions in local Dutch coordinates (after: Polman, J. et al., 1996).

3. Level plane. Points positioned on the same level plane (equal potential of gravity) all exhibit equal dynamic heights irrespective of the elevation. However, two level planes at two largely different elevations do not exhibit a constant difference in orthometric heights H, due to the inhomogenities in the earth's gravity field.

Figure 3.6 gives the process scheme, converting GPS positions into local *XY* and *H*.

The GPS positions are given in 'length' ($\lambda$), 'width' ($\phi$) and 'height' (*h*) with reference to the WGS 84-ellipsoid.

In the Netherlands two local reference systems are used:

1. The Dutch height reference system, NAP. The orthometric heights *H* are given with reference to Normaal Amsterdams Peil, NAP (the Dutch MSL). The datumpoint is 1,429 m below a bronze bolt under the pavement of the Dam in Amsterdam.

2. The Dutch local datum for mapping, RD. The RD system (Rijksdriehoeksmeting) is using the Bessel-ellipsoid to calculate the *XY*

Figure 3.7. A Leica receiver
and a total station.

coordinates. The datumpoint is in an old church (O.L. Vrouwenkerk)
in Amersfoort.

Figure 3.7 shows part of the activities during a river survey using GPS,
carried out in 1993, South of Valencia in Spain.

## 3.4  INSTRUMENTS FOR TRADITIONAL POSITION FIXING

When position fixing is carried out, the following instruments can be used;
– optical instruments, like rangefinder, theodolite and sextant.
    These instruments are discussed in this section.
– electronic instruments, are not discussed in this course.
    Electronic instruments are particularly used if the distances to be meas-
    ured are too large for measurements of sufficient accuracy by optical
    instruments. Electronic instruments are not only costly but also rather
    complicated and, therefore, require skilled technicians for servicing
    and repairs. However, electronic instruments enable fast consecutive
    measurements and can be adapted to produce a direct digital output
    for processing by computer.

*Range finder*
The coincidence type range finder is an optical instrument to determine
the distance between two points by observations from one point only.

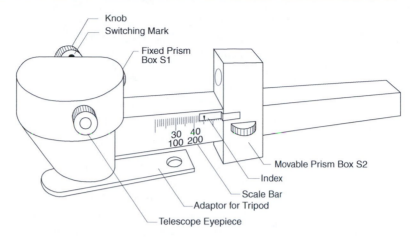

Figure 3.8. Range finder
with a shifting mirror (after:
Topcon).

All range finders are equipped with two mirrors or prisms:
–   the fixed one $S_1$ at the left hand, and
–   the rotating or shifting one $S_2$ at the right hand

Figure 3.8 shows a Topcon range finder with a shifting declination prism $S_2$ (movable prism box).

For most range finders with a shifting mirror or prism, the direct image of the measured point (coming from the fixed prism) is located in the centre of the lower half circle. The indirect image (coming from the movable prism) is located slightly beside the centre of the upper half circle. Both images can be made to coincide by moving the shifting mirror. Usage of this range finder:
–   hold the fixed prism box with the left hand and the movable prism with the right hand, keeping the scale bar horizontally
–   estimate the distance visually and switch the distance scale knob either to long distance or short distance
–   rotate the eyepiece ring until the target point is distinctly seen (depending on the user's eyes)
–   view both images in the telescope-circle, and move the shifting mirror/prism left or right, until the upper one coincides with the lower one, correctly and clearly
–   read the distance on the scale according to the distance scale
    The maximum error of this range finder:
    short distance (8–100 meter)    : 1% of the measured distance
    long distance (100–500 meter) : 3% of the measured distance
    Range finders are mostly used by holding the instrument manually, and from a moving survey boat. For measurements to be taken from a bank, a camera tripod can be used (leading to higher accuracy).

*Theodolite*
Theodolites can be used to measure vertical and horizontal angles. The most modern (and expensive) instruments are the fully digital electronic theodolites.

For position fixing in hydrometry, theodolites are used to measure horizontal angles from a fixed position at one of the banks (indirect method).

*Sextant* (Figure 3.10)
The sextant was originally designed for measuring vertical angles, such as the sun's or star's altitude from aboard a ship. However, it also serves well for the measurement of horizontal angles, specially from aboard a ship, when a theodolite cannot be used.

The measuring range is from 0°–115°, while the graduated arc holds 60°, from which the instrument got the name sextant.

Principle: Two beacons *L* and *R* (or other marks) – the angle of which between them is to be measured – are brought to coincidence in the telescope. The telescope is directed to the left hand beacon *L*, which is seen directly through the telescope. The right hand beacon is double reflected by the rotating mirror *G* and the fixed mirror *K* in front of the telescope.

The circular scale is connected with the centre of the circle. The mirror *G* rotates around a vertical axis through the centre of the circle. The telescope only serves to enlarge and clarify the view. The centre line of the telescope intersects the upper rim of the fixed mirror *K* so that the direct view of point *L* is obtained in the upper half of the telescope, while in the lower half the twice-reflected view of point *R* can be seen. When these images coincide the angel ϕ can be read on the scale with an accuracy of up to half a minute.

Figure 3.9. Images in the telescope.

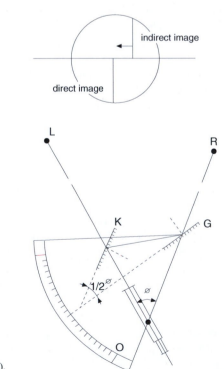

Figure 3.10. Principle of sextant (after: Jansen, 1979).

The accuracy of a sextant is small, compared to the theodolite. For position fixing inside an existing triangulation network, it gives sufficient accuracy. The advantage of the sextant is its suitability to hold the instrument manually aboard a ship.

## 3.5  EXAMPLES OF TRADITIONAL POSITION FIXING IN A STRAIGHT RANGE LINE

The very many methods to fix a position depend on:
- local circumstances (accessibility of banks, availability of skilled personnel and the required facilities)
- the specific requirements put by the surveyor and his inventiveness
- required accuracy

As example for the position fixing in a range line of a cross section, the angular method and the linear method are presented. Both methods are indirect methods.

*Angular method* (Figure 3.11)
A theodolite is set-up on one of the banks and angular measurements are taken to the boat ($\alpha$). It is also possible to take angles with a sextant from the vessel to flags in the range line and to the reference point $T(\beta)$. This method is suitable for fixing the boat's position taking soundings. The position is $VC = CT \, tg \, \alpha$ or $VC = CT \, cotg \, \beta$.

*Linear measurements* (Figures 3.12 and 3.13)
The flags $A$, $B$, $C$ and $D$ are fixed in the range line. Flag $E$ is fixed along a line perpendicular to $AD$ and with a known distance from $B$. The observer $O$ moves along a line also perpendicular to $AD$ until the flag $E$ and flag $V$

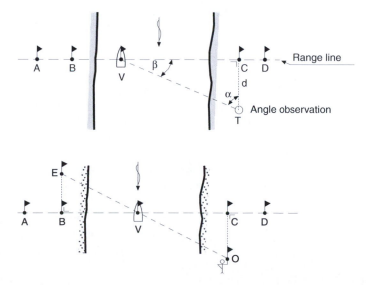

Figure 3.11. Angular method (after: Hayes, 1978).

Figure 3.12. Linear method for small rivers. (after: Hayes, 1978).

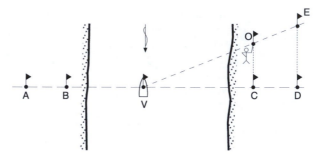

Figure 3.13. Linear method
for wide rivers (after: Hayes,
1978).

on the vessel and the observer's flag $O$ are in one line. Then the distance
$CO$ is determined and $VC$ can be computed.

$$VC = \frac{BC \times CO}{CO + BE}$$

If the river is very wide so that the flags on the opposite bank are not
clearly visible, the boat's position can be fixed at the method given in
Figure 3.13, where flag $E$ is fixed on the same bank as the observer.

$$VC = \frac{DE \times CD}{DE - CO}$$

## 3.6  EXAMPLES OF TRADITIONAL POSITION FIXING IN A FREE SYSTEM

Position fixing in a free system can be done for two purposes:
1. Sounding in very wide areas, where the setting up of straight lines
   becomes difficult or impossible.
2. Triangulation networks, for example to define the river geometry (the
   alignment and the width), when reliable maps are not available.

Many different systems are used to determine the surveyor's position
within an existing system of base points (one, two or three base points
with known coordinates).

Figure 3.14 gives a number of examples, how to determine one's
position using simple instruments like range finder, sextant and compass
(A leading line is a straight line through two base points, where the
observer is positioned in this line).
a. Two sextants, three base points
b. One sextant and a leading line, three base points
c. One range finder and a leading line, two base points
d. One sextant and a range finder, two base points
e. One sextant and a compass, two base points
f. One range finder and a compass, one base point

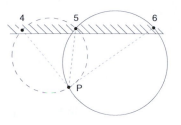

a.  Two sextants

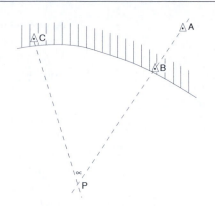

b. One sextant and a leading line

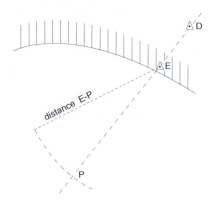

c. One range finder and a leading line

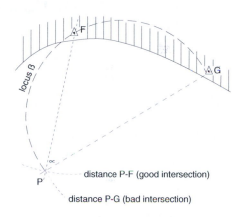

d. One sextant and a range finder

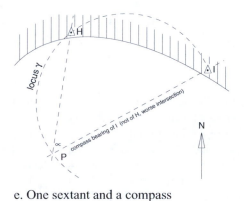

e. One sextant and a compass

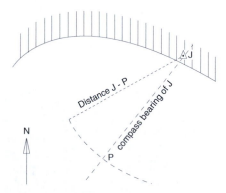

f. One range finder and a compass

Figure 3.14. Examples of position fixing in a free system (after: Nedeco, 1973).

The method to be selected depends on the following conditions:
– available instruments
– available number of base points
– inventiveness of the surveyor

## 3.7  SOUNDING AND SOUNDING INSTRUMENTS

In rivers the cross sections are sounded in transit lines perpendicular to the river's center line, and at fixed intervals spaced along the river.

Position fixing is done by direct methods (tape and range finder) or by indirect methods (linear and angular). At the same time the position fix is indicated on the echosounder recording.

In estuaries or along open coasts, soundings are taken on lines perpendicular to the bottom- contour lines in order to have the most accurate way of locating these lines. The intervals between the tracks are more or less dependent on the purpose of the survey and the scale on which data should be charted.

*Various instruments are used for sounding (depth measurements)*
The sounding rod and sounding line were formerly the conventional tools used for measuring the depth of water. The accuracy of both methods is limited and the echo sounder has found wide application instead because it is not restricted by flow velocity and depth. It is, therefore, universally more acceptable than the outdated conventional tools. It also enables a fast measurement of the depth.

*Sounding rod*
The *sounding rod* is a wooden rod graduated in centimetres, or decimetres. At the bottom end of the rod is a base plate which prevents the rod from penetrating the river bed and also serves as a weight to keep the rod in a vertical position on the bed. Sources of error are deviation of the rod from the vertical and the stagnation head related to the flow velocity, especially where there are strong currents. Therefore, the applicability of this tool in flowing water is limited.

*Sounding line*
The *sounding line* is used if the flow velocity or the water depth prevents the use of the rod. It consists of a chain or wire to which a heavy weight is attached. Tags are tied to the line to indicate metres and decimetres. In other cases (use of davit and winch) the depth is read from a counter aboard ship. Corrections have to be made to the measurements to allow for deviations from the vertical above the water surface (air line corrections) and below the water surface (wet line corrections). For this purpose the international standard ISO 3454 has been developed.

The depths which can be measured by a sounding line are limited to some 12 to 15 m. Sources of error are the penetration of the weight into the river bed and the variations from the ideal conditions (wire bending etc.). Furthermore, it may be difficult to ascertain whether the weight is actually in contact with the river bed.

### Echosounders

The *echo* or *ultrasonic sounder*, the use of which is not limited by flow velocity or depth, is the most suitable tool for sounding. It enables accurate and rapid measurement of depth, based on the propagation speed of sound in water and the measurement of the running time of the sound wave. The propagation speed of a sound wave in water depends on temperature and salinity. Increase of each of them leads to increase of $C_{water}$. Figure 3.15 gives the propagation speed of a sound wave in water as a function of the water temperature and the relative density.

The principle is as follows: a short but strong electrical pulse is produced by the signal generator, amplified and sent to the transducer which converts it into an acoustic signal which is sent to the river bed. The signal is then reflected from the bed, received by the transducer and converted into an electrical pulse, which is amplified. The time-lapse between the transmission and reception of the signal is measured and converted into the depth of water under the transducer. The depths are shown on a

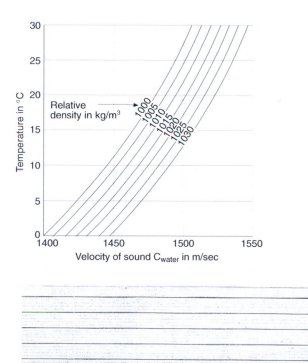

Figure 3.15. Propagation speed of a sound wave as a function of temperature and relative density.

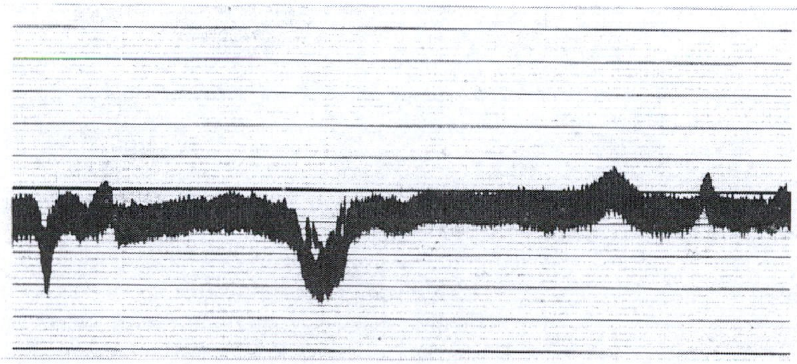

Figure 3.16. Echogram, bottom level on recording paper.

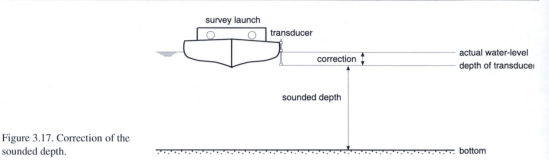

Figure 3.17. Correction of the sounded depth.

suitable indicator, recorded on a paper roll or given as a digital output. Because of the high transmission frequency, the individual depths appear on the recording paper as an almost continuous bed profile, as is shown in Figure 3.16.

The transducer is located at a certain distance (0.20 to 0.50 metres) below the water level. And as the distance between the bottom of the transducer and the reflected surface of the river bottom is measured, the draught of the transducer should be added to the recorded depth, so giving the correct water depth (See Figure 3.17). The transducer may be onboard (built in the vessel's bottom) or outboard, fixed to the ship with a bracket. Most echosounders are operated on a 12 V battery.

From time to time the echosounder must be calibrated. Besides that, at the beginning and end of each sounding day, the zero-setting of the echosounder should be checked by means of a bar check. Zero setting depends on the depth of the transducer under the water surface, which may vary from day to day. In the case of an outboard transducer, a steel plate suspended to a marked line is lowered into the water and held at a certain depth beneath the transducer. The depth recording of the testbar should correspond with the actual depth of the testbar under the water surface. This is checked at various depths. If this is not the case, an adjustment has to be made in the speed of sound to cope with the density and temperature of the water, until the echosounder reading corresponds with the actual depth of the testplate. Air bubbles under the transducer should be avoided by limiting the speed. Usual speeds are 1.50 to 2.00 m/s.

The echosounder is used for cross-sectional soundings of the river, longitudinal soundings (talweg) and local complete soundings of a part of the river, a bay or a lake.
The paper speed can be adapted to the different types of soundings:
– river cross-sections: the highest paper speed
– longitudinal soundings; a lower paper speed

The echosounders used in hydrological surveys have in most cases frequencies of 30 kHz or/and 210 kHz.

If soft mud layers are expected or the survey is intended as a predredging survey or a post-dredging survey, a low frequency echosounder should be used in order to detect layers and sedimentation or degradation, as the sound pulse of a low frequency echosounder penetrates more deeper in the bottom. A high frequency echosounder, however, gives only a recording of the top of the bottom regardless the bottom composition. By using both frequencies simultaneously the multiple layers in the riverbed are detected.

*Determination of water level*

For conversion of the measured depth into a level with respect to the datum plane, gauge readings have to be made during the soundings. Usually, the water level in the sounding area is determined by linear interpolation between two readings at existing gauging stations. If such interpolation introduces an error exceeding a few centimetres, a temporary gauge should be installed close to the sounding area. As the error due to linear interpolation over time should not be more than a few centimetres, the frequency of the readings depends on the rate of change of the water level. Therefore, during periods of rapidly changing water levels, frequent readings are required.

## 3.8  DATA PROCESSING

Each of the components of a sounding procedure, i.e. the determination of the location of sounded points, of the water level and of the depth measured with respect to the water level, will result in a number of data. To arrive at a suitable presentation of the results of the soundings, these data have to be processed.

When the sounding takes place in cross sections, the data for the determination of the location of a sounded point consist of the number of the cross section, the location of the reference mark in the cross section and the distance of the sounded point to the reference mark. When the free sounding system is applied, the data consist of, for instance, the coordinates of the base points used and two angles measured simultaneously. The location can then be determined either mathematically or graphically.

The water level at the sounding point is determined using records from one or two gauges. From these data, as well as the gauge-zeros and their locations, the required water level is obtained by linear interpolation over distance and, if necessary, over time. The time link between soundings and water level is essential. The water level and the sounded depth, obtained from an echogram or a list of depths measured, are then used to determine the elevation of the sounded points of the river bed with respect to a horizontal datum plane (MSL).

All components of the processing are rather simple and can easily be done manually. However, if the extent of the work justifies it, the entire processing, or parts of it, can be programmed for computer. Other data such as gauge readings and time are limited in number and can easily be brought into digital form.

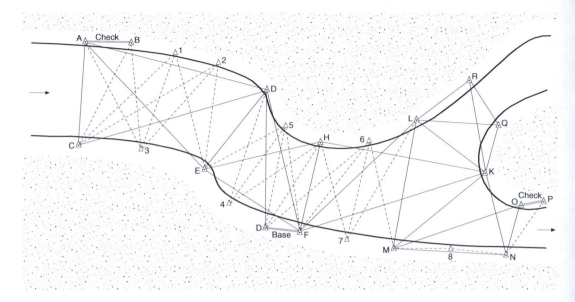

A, B, C, D etc. main triangulation beacons
1, 2, 3, 4, etc. auxiliary beacons used for local positioning
G-F: main base, A-B and O-P: second and third base
L-R, K-Q, M-8 and 8-N as well as the bases, sighting is possible. In other cases sighting might be obstructed by bushes or trees
and if nessesary sighting paths should be cut.

Figure 3.18. Example of a triangulation network to get the river alignment (after: Nedeco, 1973).

Figure 3.18 is an example of a triangulation network, to get the alignment of a river stretch.

The results of soundings are usually presented in the form of longitudinal- or cross sections of the sounded area or as sounding charts, which show the sounded points and their depths with respect to the datum or reference plane.

For soundings taken in a river according to cross sections, the water depths related to the reference plane for each point in the cross section can be plotted on mm-paper and a cross sectional profile can be drawn (see Figure 3.19). If so required, a complete contour chart of the river can be made derived from the cross sectional profiles.

On the sounding chart, contour lines indicating equal depths can be drawn. This work requires experience and understanding of the draughtsman, and it can be improved by occasional detailed measurements between the sections which are normally sounded.

Figure 3.20 shows a sounding chart with contour lines.
The spacing of sounding lines (sounding tracks or transit lines) for making a sounding chart depends on:
– the water depth and river alignment,
– the shape of the river bottom,
– the purpose of the survey.

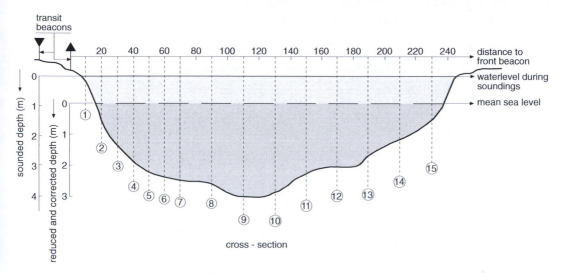

Figure 3.19. Example of a cross sectional profile. (after: Hayes, 1978).

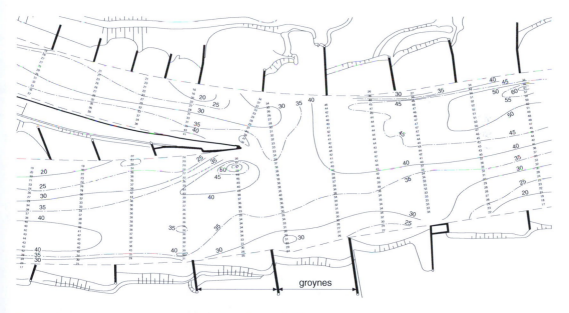

Figure 3.20. Sounding chart with contour lines (after: Jansen, 1979).

In most cases a spacing of 100 metres is acceptable for river charts, depending on the river width and the river alignment.

Developments in progress are:
– radar for bathymetry
– air borne lasers

## 3.9  INTERNATIONAL STANDARDS

The following standards are available:
– ISO 3454    Direct depth sounding and suspension equipment (rods, cables and weights)
– ISO 4366    Echosounders for water depth measurements
– ISO 6420    Position fixing equipment for hydrometric boats (description and selection of instruments)

CHAPTER 4

# Discharge measurements

## 4.1 INTRODUCTION

*Classification of flows*
Flows can be classified by two parameters: time and distance
1. the first subdivision is based on consideration of the time scale, classifying flows as either steady or unsteady,
2. the second subdivision relates to the scale of distance, classifying flows as either uniform or non-uniform.

The majority of flows in open channels will fall into one of the classifications listed below.

*Steady or permanent uniform flow.* The discharge is constant with time, and the cross-section through which the flow passes is of constant area. A typical example is that of constant flow through a long irrigation canal with uniform cross-section, free from back water effects.

*Steady non-uniform flow.* The discharge is constant with time, but the cross-sectional area varies with distance. Examples are flow with constant discharge in a river (the cross-section of a river usually varies from point to point).

*Unsteady uniform flow.* The cross-section is constant, but the discharge varies with time. This is a complex flow pattern. An example is that of a pressure surge in a long straight pipe of uniform diameter.

*Unsteady non-uniform flow*, in which the cross-section and discharge both vary with time and distance. This is typified by the passage of a flood wave in a natural channel, and is the most complex flow to analyse.

Discharge measurements can be carried out for all four flow types.

Finally, each of the four flow types can be subcritical, critical or supercritical, depending on the value of the Froude number

$$Fr = \overline{v} / \sqrt{g \cdot A/B_s} \tag{4.1}$$

in which

$Fr$ = the Froude number (–)

$\overline{v}$ = average flow velocity in the cross-section (m/s)
$A$ = cross-sectional area (m$^2$)
$B_s$ = width at the water level

For wide rivers $A/B_s \approx d$, where $d$ is the average waterdepth, resulting in $Fr = \overline{v}/\sqrt{gd}$ .

| | | |
|---|---|---|
| subcritical flow | $Fr < 1$ | mild bedslopes, moderate flow velocities |
| critical | $Fr = 1$ | |
| supercritical | $Fr > 1$ | steep bedslopes, high flow velocities |

*Basic concept of discharge*
In open channels (artificial canals and natural rivers) the discharge $Q$ in any cross-section and at any moment is given by:

$$Q = \overline{v} \cdot A \tag{4.2}$$

Thus the discharge can be determined if the area $A$ is known (or measured) and the average velocity $\overline{v}$ is calculated on the basis of velocity measurements carried out in that particular section.

In this Chapter the determination of the discharge $Q$ in open channels is emphasized.

### 4.1.1 *Purpose*

Flow measurement in open channels may serve several purposes:
–  basic information on river flow for the design of diversion dams and reservoirs and for putting up bilateral agreements where rivers are forming or crossing the national frontiers;
–  distribution of irrigation water
–  basic information for charging industries and treatment plants draining away their polluted or purified water in public water courses
–  basic information for other water users, such as navigation
–  water management in urban and rural areas: storage of fresh water and removal of rainwater excess
–  reliable statistics to be based on long term monitoring (stochastic behaviour of the rainfall-runoff system)
In most cases incidental measurements are not satisfactory, but continuous measurements are indispensable.

### 4.1.2 *Relation discharge – bed roughness*

In open channels a relation exists between the water stages and the discharges and, generally, it is known that higher stages correspond to higher discharges.

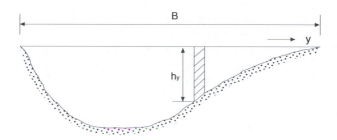

Figure 4.1. Discharge cross section.

Expressing the discharge of a river as $Q = \bar{v} \cdot A$ and inserting the equation of Chézy ($v = C\sqrt{R \cdot S}$), the discharge formula is derived (see Figure 4.1) thus:

$$Q = \int_o^B C\sqrt{R \cdot S} \cdot h_y dB \qquad (4.3)$$

in which
$Q$ = discharge (m³/s)
$C$ = Chézy coefficient for the total bed roughness (m$^{\frac{1}{2}}$/s)
$S$ = gradient of the energy level along the river; in practice, approximated by the water level gradient (–)
$R$ = hydraulic radius, defined as the wetted area of the cross-section, divided by the wetted perimeter, $R = A/P$ (m)
$h_y$ = depth of the river at allocation $y$ (m)
$B$ = width of the river (m)

As for most rivers the width is very large compared with the depth, the hydraulic radius can often be replaced by the depth and Equation (4.3) then yields:

$$Q = \int_o^B C\sqrt{S} \, h_y^{\frac{3}{2}} \, dB \qquad (4.4)$$

In hydrometric surveys it is common knowledge that discharges have to be measured quite extensively. However, when studying Equation (4.4) it appears that with the known characteristics of a certain cross-section (the relation between the depth $h_y$ and the width) and an estimate of the bed roughness ($C$), only measuring the water level gradient ($S$) will be sufficient to determine the total discharge of the cross-section. This procedure may well be adopted for fixed wall channels with known roughness parameters.

In alluvial channels with continuously changing bed forms it is quite difficult to estimate the bed roughness coefficient $C$. A change in the water level will also result in a change in the bed level, but the latter may locally differ considerably from the former (particularly after flood periods). Consequently, only by measuring the total discharge of a cross-section and the local water level gradient will it be possible to compute afterwards the bed roughness coefficient. As this coefficient must be used again in the sediment transport formulae, the necessity for extensive discharge measurements in alluvial channels is clearly demonstrated.

## 4.2 VARIOUS METHODS OF DISCHARGE MEASUREMENTS

For the determination of the discharge in an open channel – rivers and artificial canals – various methods can be applied.

The majority of the flow measuring methods – but not all of them – are based on the simple concept: $Q = \bar{v} \cdot A$.

We can identify roughly nine methods, of which three *single measurement methods* and six *methods of continuous measurements*. Single measurements are carried out occasionally or for a short period, often to calibrate one of the methods of continuous measurements.

The following methods can be understood as *single measurements*:

1. *Velocity area method:* the area of the cross-section is determined from soundings, and flow velocities are measured using current-meters, electro-magnetic sensors, or floats. The velocity area method is also referred to as the current-meter method.
   The mean flow velocity is deduced from velocities measured at points distributed systematically over the cross-section. The discharge is then defined as $Q = \bar{v} \cdot A$. In principle, the measurements should be made under steady conditions. This requirement may restrict application of the conventional velocity area method, for instance in wide rivers where, due to the number of measurements involved, the time needed to take a complete set of measurements may be too long.
   The conventional velocity area method is discussed in Section 4.3. When this method cannot be used (wide rivers, unsteady flow conditions), the moving boat method using the conventional current-meters can be applied as an alternative velocity area method, enabling fast measurement, which is discussed in Section 4.3.10. In addition ADCP measurements are discussed in Section 4.3.12.
2. *Slope area method*: from measurements of the water surface slope $S$, the cross-sectional area $A$ and the hydraulic radius $R$ and by estimating a roughness coefficient for the channel boundaries, the discharge can be calculated using the Manning equation or the Chézy equation. The Manning equation reads:

$$Q = \frac{1}{n} \cdot R^{\frac{2}{3}} \cdot S^{\frac{1}{2}} \cdot A \qquad (4.5)$$

in which $n$ is the Manning coefficient.

This much less accurate slope area method can be used to determine a peak discharge after floods have receded. If insufficient gauge readings are available to calculate $S$, the flood mark which is left after the passage of the flood can be used for defining the slope $S$.
The method is discussed in Section 4.4.

3. *Dilution methods:* a suitable selected tracer is injected at the upstream section of the measurement reach of the stream with the Mariotte bottle. In a downstream sampling section, samples are taken at regular time intervals. The discharge of the stream can be calculated from the concentration of tracer injected at the upstream section of the reach and the concentration measured at the downstream end of the reach where the tracer should be uniformly distributed throughout the cross-section. Generally, this condition restricts the applicability of dilution methods to mountain streams and torrents with a high degree of turbulence. In such cases, however, fairly accurate results can be obtained. The maximum discharge to be measured with a Mariotte bottle $V = 25$ liters, will be about 0.5 m$^3$/s. Discharges > 0.5 m$^3$/s can be measured using larger bottles, volumetric pumps or constant level tanks. The method is extensively explained in Section 4.5.

The following methods can be understood *as continuous measurements*:
1. *Stage discharge method*: Once a unique relation has been established between water level and discharge by one of the single measurement methods, discharges are derived from the continuously measured water levels. Section 4.6.
2. *Slope stage discharge methods*: First a relation must be set up between water level, water surface slope and discharge based upon the Manning equation or the Chézy equation and calibrated by one of the single measurement methods. After this, discharges are derived from two water levels that are measured continuously. Section 4.7.
3. *Acoustic method*: Discharges are calculated from measurements of both the flow velocity and the water level. The velocity is computed from the difference in running time of a sound wave which is transmitted diagonally across the channel in upstream and downstream direction. Section 4.8.
4. *Electro-magnetic method*: The flow velocity is determined by measuring the voltage induced by a moving conductor (streamflow) in a magnetic field. Section 4.9.
5. *Pumping stations*: For any pumping station a relation can be established between the discharge and the total loss of head, supported by calibration with one of the single measurement methods. By counting the pumping hours, the total volume of water can be calculated. Section 4.10.
6. *Flow measurement structures*: Discharges are derived from measurements of the upstream water level which is continuously measured at a certain distance upstream of the structure (weirs, gates and flumes).

Table 4.1. Overview of discharge measurement methods.

| Method | | Frequency of measurements | |
|---|---|---|---|
| | | Incidentally | Continuously |
| Velocity area method | (4.3) | x | - |
| Slope area method | (4.4) | x | - |
| Dilution method | (4.5) | x | Sometimes |
| Stage discharge method | (4.6) | - | x |
| Slope stage discharge method | (4.7) | - | x |
| Acoustic method | (4.8) | Sometimes | x |
| Electro magnetic method | (4.9) | - | x |
| Pumping stations | (4.10) | - | x |
| Flow measurement structures | (6) | Sometimes | x |

The use of flow measuring structures is discussed in Section 4.11 and Chapter 6.

The above methods all have their specific applications. When a selection should be made to employ one of them, the following considerations are usually brought in discussion:
– do we want single measurements or continuous measurements?
– hydraulic conditions (channel stability, type of flow, sediment contents, etc.)
– what is the required accuracy?
– do we have skilled surveyors available?
– do we have sufficient head available? (pumping stations and flow measurement structures)
– what are the costs of installation and operation?

When judging the applicability of instruments, aspects to be considered are:
– the availability of power supply required for operating the instrument
– the availability of spare parts
– the possibility of small repairs by unskilled personnel and the presence of a nearby workshop or dealer for more complicated repairs
– the necessity of calibration and the method of calibration (current meters)
– the dimensions and the weight of the instrument with regard to transport and the use of survey boats, winches, etc.

The International Standard ISO 8363 'General guidelines for the selection of methods' gives the restricting conditions when selecting the most suitable method (Section 4.12).

## 4.3  VELOCITY AREA METHOD

### 4.3.1  *Introduction*

The velocity area method is directly based on the concept $Q = \bar{v} \cdot A$
where
$Q$ = discharge (m³/s)
$\bar{v}$ = mean velocity in the cross-section (m/s)
$A$ = cross-sectional area (m²)
   The cross-section should be selected so, that the main flow direction
is perpendicular to that section, as much as possible.

The cross-sectional area is calculated from the measured width in the
section, and the water depths in a number of verticals. The width is measured with a tape or by applying one of the position fixing methods, discussed in Sections 3.3 and 3.4. The depths are sounded in a number of
$m$ verticals, equidistantly spaced between both banks. The number of
verticals ($m$) must be such, that the shape of the cross-sectional profile is
described accurately, varying from $m = 5$ for regular and small sections
until $m = 15$ for irregular and wide sections.
   It is common use in hydrometry to measure all parameters, starting
from the left bank.
   Figure 4.2 shows a cross-sectional profile. The reference point *RP* is
located at the left bank. Water depths are measured in $m = 7$ verticals.
   The flow velocity distribution in the cross-section can only be defined
by measuring the velocities in the pre-selected verticals. It is impossible
to measure the mean velocity $\bar{v}$ directly.

### 4.3.2  *Velocity distribution in the cross-section*

*Velocity distribution in a vertical*

Figure 4.2. Example of a cross section.

Steady or permanent flow is defined as flow with $dQ/dt = 0$ (and thus
$dv/dt = 0$). In all points of a cross-section the velocity does not change

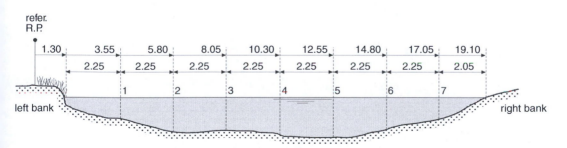

width at water level : 19.10 - 1.30 = 17.80  m

selected: 8 panels  ➝  7 verticals  ➝  equidistance: 17.80 / 8 = 2.225 m.
                in practice   : 7 x 2.25 + 1 x 2.05 m.
                            ( 6 x 2.25 + 2 x 2.15 m.)

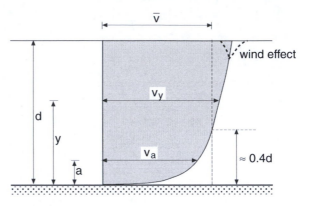

Figure 4.3. Parabolic velocity
distribution in a vertical.

with time at all (permanent flow) or does only change very gradually
(dv/d$t$ ≈ 0), (quasi-permanent flow).

Due to bottom friction, in general the velocity near the bottom of an
open channel will be less than near the water surface. Near the embank-
ments the roughness of the side slopes will have influence on the velocity
distribution in the channel.

In a vertical the velocity distribution depends on whether the flow is laminar or turbulent, which can be seen from the Reynolds number

$$Re = \frac{v \cdot D}{v}$$  (4.6)

where
$Re$    the Reynolds number (–)
$D$     water depth (m)
$v$     kinematic viscosity (m²/s)
laminar flow            $Re < 400$          } rarely found in
transitional flow     $400 < Re < 800$   } natural water courses

turbulent flow          $Re > 800$          normally in
                                            natural water courses

*Example:*

v ≥ 0.01 m/s          }
D ≥ 1.0 m and       } $Re = \dfrac{v \cdot D}{v} = \dfrac{10^{-2} \cdot 1}{10^{-6}} = 10^4 \rightarrow$ turbulent flo\
$v$ ≈ 1.10$^{-6}$ m²/s  }

The velocity distribution in a vertical – for turbulent flow – is assumed to
be parabolic, as shown in Figure 4.3.

The parabolic velocity distribution is expressed by Equation (4.7)

$$\frac{v_y}{v_a} = \left(\frac{y}{a}\right)^{\frac{1}{n}} \tag{4.7}$$

where
$y$ and $a$  = arbitrarily chosen distances from the bottom
$v_y$ and $v_a$ = flow velocities in $y$ and $a$
$n$          = power coefficient

For natural rivers $n$ lies between 5 and 7, but it may vary over a wider range, depending on the hydraulic resistance: a low $n$-value applies to course beds, and a high $n$-value applies to smooth beds. The correct $n$-value can be calculated from Equation (4.8):

$$n = 1.77 + f \cdot C \tag{4.8}$$

where
$f$  = is a conversion factor. $f = 0.098$ ($\text{m}^{-0.5} \cdot \text{s}$)
$C$ = Chezy's bed-roughness coefficient ($\text{m}^{0.5} \cdot \text{s}^{-1}$)

The mean velocity $\overline{v}$ in the vertical is found approximately at a distance $y_{\overline{v}} = 0.4\,d$ from the bottom ($d$ is the total water depth). In fact, this depth varies slightly with the $n$-value, and thus with the bed-roughness. The depth $y_{\overline{v}}$ as a function of $n$ can be derived from Equation (4.9).

$$y_{\overline{v}} = \left(\frac{n}{n+1}\right)^{n} \cdot d \tag{4.9}$$

Table 4.2 gives some values $y_{\overline{v}}$.

Using the Equations (4.7) and (4.9), the mean velocity $\overline{v}$ can now be calculated from a measured velocity $v_y$ at any depth $y$ (related to the bottom) as shown in Table 4.3.

Table 4.2. Depth $y_{\overline{v}}$ as a function of $n$.

| n | $y_{\overline{v}}$ |
|---|---|
| 3 (extremely rough) | $0.422\,d$ |
| 5 | $0.402\,d$ |
| 6 (normal value) | $0.397\,d$ |
| 7 | $0.393\,d$ |
| 10 (extremely smooth) | $0.386\,d$ |

Table 4.3. Mean velocity $\bar{v}$ in the vertical, related to $v_y$, $y$ and $n$.

| Depth $y$ (from bottom) | Relative velocity $\bar{v}/v_y$ for different $n$-values | | | | | Depth from water surface (ISO) |
|---|---|---|---|---|---|---|
| | 3 | 5 | 6 | 7 | 10 | |
| 0.2 $d$ | 1.28 | 1.15 | 1.12 | 1.10 | 1.07 | 0.8 $d$ |
| 0.4 $d$ | 1.02 | 1.00 | 1.00 | 1.00 | 1.00 | 0.6 $d$ |
| 0.5 $d$ | 0.94 | 0.96 | 0.96 | 0.97 | 0.97 | 0.5 $d$ |
| 0.8 $d$ | 0.81 | 0.87 | 0.89 | 0.90 | 0.93 | 0.2 $d$ |
| $d$ | 0.75 | 0.83 | 0.86 | 0.88 | 0.91 | 0 |

*Conclusions*
1. the mean velocity $\bar{v}$ in the vertical is found at a depth 0.4 $d$ from the bottom (or 0.6 $d$ from the water surface), provided $n \geq 5$.
2. the variation in velocity $v_y$ in the vertical is fairly strong for a rough bed, and fairly small for a smooth bed.

In every point of the vertical 'the velocity' in fact is also an average. Because of the turbulent character of the flow in any point the flow fluctuates around its average value, as shown in Figure 4.4.

Thus, when measuring a point velocity $v_y$ with a current-meter, it is recommended to measure during a minimum period of 30 to 60 seconds, in order to find a reliable average velocity
– high velocities    $t = 30$ to 50 seconds
– low velocities    $t = 60$ to 100 seconds

*Velocity distribution in the cross-section*
The velocity distribution in a cross-section strongly depends on the shape of the section. Also the configuration (bends!) of the channel upstream of a cross-section is important. Examples are shown in Figure 4.5.
   It can be seen that to obtain the mean value of the velocity for a cross-section, both the vertical and horizontal velocity distribution should be taken into account. This means that both in vertical and in horizontal direction a number of measuring points are necessary to have a sufficiently significant 'sample'.

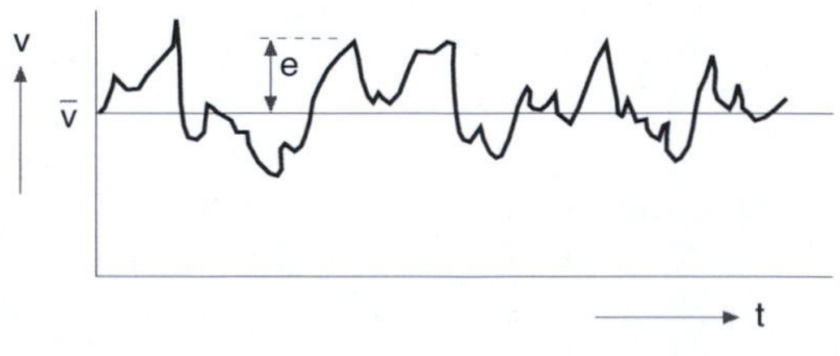

Figure 4.4. Variation of velocity with time in turbulent flow.

v = actual velocity
$\bar{v}$ = time averaged velocity
e = stochastic component

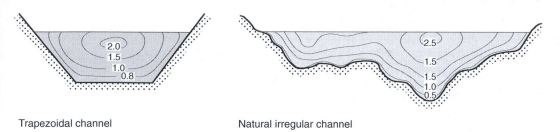

Trapezoidal channel          Natural irregular channel

Figure 4.5. Examples of velocity distribution in a cross section.

*Question*: how many verticals are required and how many measuring points per vertical?

The *answer* depends on:

1. The required *accuracy* (the more points used, the more accurate $\overline{v}$ will be)
2. *Economical considerations* (more measuring points require more time)
3. Practical circumstances, e.g. unsteady flow conditions, the required time for an extensive measurement is not available (see also the relevant ISO standards)

With respect to accuracies that can be obtained, a study was done by a Technical Committee of the International Organization for Standardization (ISO). The results of an analysis of a great number of Q-measurements, done in rivers in USA, UK, The Netherlands and India, and some conclusions of this study were summarized in WL Delft Hydraulics Publication No. 106, as follows:

'For approaching the mean velocity in each vertical a number of points in the vertical can be taken. As long as the velocity distribution is 'normal' (a more or less parabolic velocity distribution), the number of points in the vertical ($n$) is of less importance than the number of verticals ($m$) between both banks.' Figures 4.6 and 4.7 show the results of the ISO-investigation (based on measurements in various rivers):

Figure 4.6 shows the standard deviation in $\overline{v}$ as a function of the number of points $n$ in the vertical.

It can be concluded that taking $n > 3$ is hardly increasing the accuracy. In Section 4.3.8 this standard deviation is referred to as $\sigma_{f+s}$ (error type I and II).

Figure 4.7 gives the standard deviation in $\overline{v}$ as a function of the number of verticals $m$ between both banks. In Section 4.3.8 this standard deviation is referred to as $\sigma_{d+h}$ (error type III).

Similar to the number of points in each vertical, the number of verticals show to have a large influence on the final error.

This is not surprising, as most rivers do have relatively high width/depth ratios. In practice this is not always realized by engineers, partly due to the understandable habit to draw river cross-sections on strongly distorted scales.

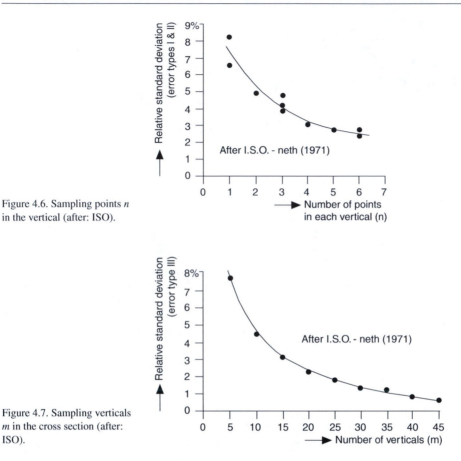

Figure 4.6. Sampling points *n* in the vertical (after: ISO).

Figure 4.7. Sampling verticals *m* in the cross section (after: ISO).

Figure 4.8 is an example of an undistorted cross-section: the horizontal scale and the vertical scale are the same.

The results of the ISO investigation show clearly that a proper set-up of the velocity area method may lead to an increased accuracy. It is also a plea for the moving boat method, where the horizontal velocity distribution is measured in an accurate way.

### 4.3.3 *Selection of site*

Figure 4.8. Cross-section of a river (horizontal and vertical scale the same).

The site of the cross-section, no matter whether it is located in tidal or non-tidal rivers, should comply with the following conditions:

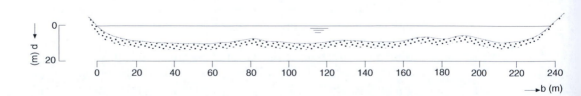

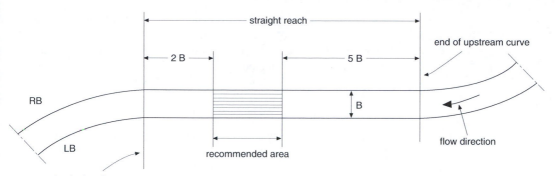

Figure 4.9. Recommended
site in a curved river.

– the cross-section should preferable be in a straight reach which is of
   uniform profile (see Figure 4.9).
– the water depth in the selected cross-section should be sufficient to
   provide effective immersion of current-meters or floats whichever are
   to be used;
– the orientation of the reach should be such that the direction of flow is
   as closely as possible perpendicular to that of the prevailing wind;
– sites at which vortex or backwater flow or deadlines tend to develop
   should be avoided;
   It is recommended to select the site upstream of a control (see Section
   4.6.2)
– all discharges should be contained within a defined channel with stable boundaries (banks) and well defined geometrical dimensions;
– the selective cross-section should be marked by means of transit beacons at both banks;
– a staff gauge or water level recorder should be established at the
   cross-section;
– the site shall be accessible to regular visits of the surveyor.

Figure 4.10. Site selection at a
confluence.

– be situated in a river stretch that is stable. (If the section or the river
   downstream thereof is unstable, changing conditions will change the
   *Q-h* relation);

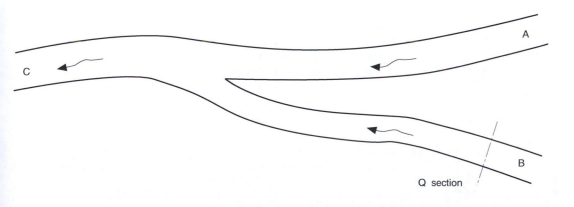

– it must be avoided to have the site close to confluences to prevent backwater effects.

Figure 4.10 shows a confluence. In the indicated Q-section in river *B* the rating curve may be effected by conditions in river *A*.

### 4.3.4 *Instruments to measure point flow velocities*

Various instruments have been developed to measure flow velocities in open channels:
– horizontal axis propeller current-meters
– vertical axis cup-type current-meters
– electro magnetic flow velocity sensors
– floats
– pendulum type meters
– velocity head rod
– pitot tube
– doppler ultrasonic velocity sensors, measuring point velocities or profile velocities (ADCP).

The first four methods are widely used, and are still fairly popular instruments; the remaining methods are less universally used.

The use of a particular meter depends on local circumstances, such as familiarity, required accuracy and budget.

All the various meters measure point velocities: velocities in a particular location in a cross section, except the ADCP (Section 4.3.12).

*Propeller current-meter*

The meter works on the principle that a horizontal axis propeller of certain pitch is turned by the water particles passing along. Each revolution of the propeller corresponds with a travelling distance of a water particle equal to the pitch of the propeller.

The number of revolutions, measured during a certain time lapse then gives the velocity, according to the calibration formula of the propeller. The general calibration equation reads as follows:

$$v = K \cdot n + \Delta \tag{4.10}$$

where
$v$ = flow velocity (m/s)
$n$ = propeller revolutions ($s^{-1}$)
$K$ = hydraulic pitch of the propeller (m)
    (determined by test runs)
$\Delta$ = characteristic meter constant (m/s)
    (determined by test runs)

The number of revolutions – the propeller pulses – are measured with a counter set. In most cases the pulses can be counted in preset time intervals of 30, 40, 50, 60 and 100 seconds. Zero setting is done by a press button.

Mainly two types of propeller current-meters are used:

1. the small or mini current-meter for use in hydraulic laboratories, and small channels. This meter is directly fixed to a rod.
Common dimensions:
    propeller diameter 30 and 50 mm
    propeller pitch 50, 100, 250 and 500 mm
    length of rods 1 or 2 or 3 metres.
2. the universal current-meter for use in rivers and deep channels. The meter is used on a rod in creeks or in shallow rivers with low flow velocities, or with a cable, suspended in larger rivers. In the latter case the measuring process is carried out with a single drum winch from a bridge or a boat, or by means of a cable way system (Section 4.3.5).
Common dimensions:
    propeller diameter 80 and 125 mm
    propeller pitch 125, 250, 500 and 1000 mm
    length of rods 3, 4, 5 or 6 metres

For low flow velocities, a propeller with a small pitch is used, for high flow velocities a propeller with a large pitch is selected.

Figure 4.11 shows the universal propeller current-meter on a rod.

The current-meter must be placed in the direction of flow (propeller in upstream direction).

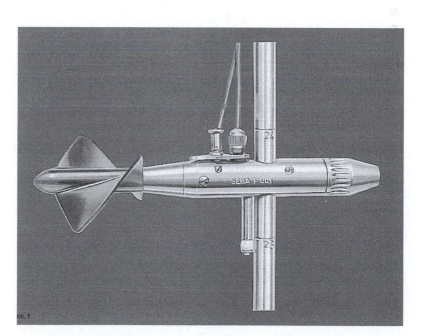

Figure 4.11. The universal propeller current-meter on a rod (after: Seba Hydrometry).

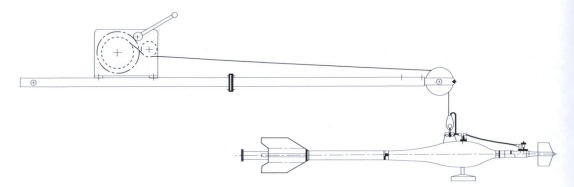

Figure 4.12. A cable-suspended universal propeller current-meter (after: Seba Hydrometry).

Using the meter on a rod, it is recommended to provide the rod with a bottom plate, to prevent the rod will penetrate in the river bed. In addition it is recommended to attach a type of telltale on top of the rod, parallel to the meter body, so as to make sure the current-meter be in the right direction.

Using the meter with a cable in a suspended way, the meter body has a heavy weight so as to keep the cable in a more or less vertical position. Sinker weights of 5, 10, 25, 50 or 100 kgs can be used (the higher the flow velocity, the more weight is needed). However, if the cable inclination with the vertical becomes more than 15°, then the depth error must be corrected according to the international standard ISO/TR 9209: Determination of the wet-line correction. Cable-suspended meters are provided with a tail and a direction vane, to get the meter in the direction of flow. Figure 4.12 shows a cable-suspended current-meter with a single drum.

The material of the propeller is anodized aluminium, brass or impact resistant plastic.

The minimum velocity, to be measured with a propeller current-meter, is v = 0.03 to 0.06 m/s, depending on the diameter and the pitch of the propeller. The error in the measured flow velocity depends strongly on its magnitude: high velocities are measured with a low error, and low velocities are measured with a relative larger error. The latter error can be reduced by taking a longer measuring period, $t > 60$ sec.

Propeller current-meters are excellent tools for fast and accurate work, provided careful maintenance is done and great care is exercised when using the instrument and on the condition that the calibration of the propellers is checked regularly.

*Cup-type current-meters*
A cup-type current-meter consists of a rotor, constructed of a number of conical cups fixed at equal angles to a frame mounted on a vertical shaft (comparable with anemometers for the measurement of wind speed).

The meter can be used both on a rod as well as cable-suspended.

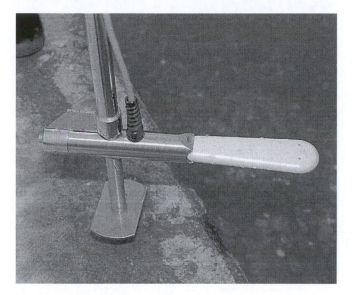

Figure 4.13. The electro-magnetic sensor (after: Ott, Flowsensor).

*Electro magnetic flow velocity sensors (EMS)*
The principle of operation of the sensor depends on Faradays Law, first discovered in 1832: 'A conductor moving in a magnetic field produces a voltage'. Similarly, if water (a conductor) moves in a magnetic field, a voltage is produced that is linearly proportional to the water flow velocity. Practical use of this concept has been made in pipe-flow metering and as ships speedometers.

Now the magnetic induction principle is also used to measure point velocities for open channel discharge measurements, so creating an alternative instrument for the conventional rotating current-meters (propeller and cup type).

Figure 4.13 shows the sensor on a rod.

Faradays Law: $E = B \cdot b \cdot v$ (4.11)

where
$E$ = measured voltage
$B$ = strength of magnetic flux
$b$ = length of the conductor (distance between both electrodes)
$v$ = flow velocity

The output voltage in the sensor is taken from the two special electrodes, it is pre-amplified and directly displayed in a physical unit (m/s).

The electro-magnetic flow velocity sensor is an adequate instrument for flow measurement in shallow water (use on a rod).

The minimum required conductivity of the measuring medium is 5 μS (the conductivity of clean fresh water is about 50 μS).

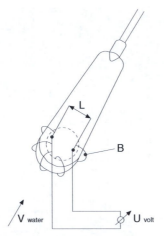

Figure 4.14. Schematic
diagram of Faradays Law
(after: Ott, Flowsensor).

The power supply comes from 10 pieces of small cells of 1.5 V each,
sufficient for approximately 30 operating hours.

The weight of the sensor is 0.5 kg.

The velocity indicator has an operating keyboard, making averaging
intervals selectable in steps between 0 and 60 seconds. The direct readout
in m/s is indicated in the display.

*Floats*

Floats are the simplest tools for measurement of flow velocity. The time
it takes the float to drift over a known distance between two previously
fixed transversal lines is a measure of the flow velocity. Measurement of
discharge by the float method is discussed in section 4.3.11.

*Pendulum-type meters*

The pendulum current-meter is used to measure the flow velocities in a
shallow river or canal, and is based on the principle that a metal or plastic
body, suspended by a thin wire from a measuring device, is moved by the
current out of a position vertically below its point of suspension. With the
help of calibration curves the angles read by the measuring device can
be translated into velocities. Corrections have to be made for the bending
of the wire. Several bodies of different shape and weight belong to the
current-meter set, each to be used in a matching range of velocities.

Figure 4.15 shows the so-called KLM pendulum current-meter, which
can easily be hand-held during the measurements for a total range of
0–2.0 m/s.

Pendulum type meters can be easily repaired and maintained in the
field, and only a relatively cheap part (the body) can be lost. The instruments give a true recording of the water movement, and the observer is
warned immediately if something is going wrong: for instance, if the
survey vessel is moving due to wind (or if the body is lost).

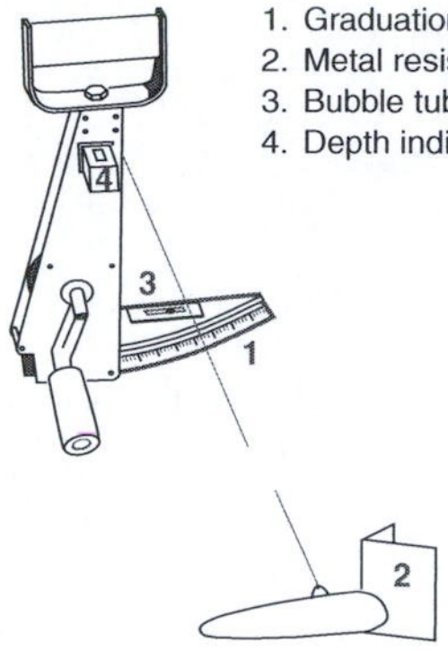

1. Graduation for reading the vertical angle
2. Metal resistance buoy
3. Bubble tube
4. Depth indicator

Figure 4.15. A typical pendulum current-meter (after: Nedeco, 1973).

Some meters – like the one of Figure 4.15 – only measure the flow velocity, other types can also read a horizontal angle, indicating the direction of the flow (in tidal areas).

*Velocity head rod*
A graded rod with a sharp longitudinal edge on one side and a blunt edge along the opposite side is placed in a stream. The rod is vertically lowered on to the stream bed with the sharp edge pointing upstream. The depth $h$ is read at the sharp edge. To measure the flow velocity, the rod is turned around ($180°$) with the blunt edge facing upstream. The resulting standing wave that builds up against the blunt edge above water level represents the average velocity head along the vertical plane in which the rod stands. The reading taken on this side represents the energy head $H$ (flow depth plus velocity head). The flow velocity is now calculated as follows:

$$v = \sqrt{2g(H-h)} \qquad (4.12)$$

where $H - h$ = velocity head
The instrument is suitable to measure high velocities $v \geq 1$ m/s, and should only be used for approximate measurements.
Figure 4.16 gives a view of the velocity head rod.

Figure 4.16. A graded velocity head rod.

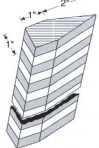

*Pitot tube*
This instrument is a well-known device in hydraulic laboratories. It is suitable for measuring high velocities, by measuring the piezometric head $h$ and the energy head $H$ in a certain location.
The velocity is then calculated as follows:

$$v = \sqrt{2g(H-h)}$$

The instrument is suitable for measuring high velocities $v \geq 1$ m/s. However, if the Pitot tube is used in combination with a differential pressure transducer, the pressure difference $H - h$ will be recorded accurately, resulting in accurate values even for low velocities.

Table 4.4. Summary of instruments.

| Instruments | Application |
|---|---|
| Propeller current-meter | Most universally used |
| Cup type current-meter | Widely used |
| | on a rod and cable suspended |
| Electro magnetic sensor | Shallow water, used on a rod |
| Surface floats | Excessive high and low velocities, less accurate |
| Pendulum type meter | Rarely used, a bit old fashioned |
| Velocity head rod | Rough estimate of velocities, inaccurate for low velocities |
| Pitot tube | Application in hydraulic laboratories |
| Doppler velocity sensor | see Section 4.3.12 ADCP, modern system |

*Doppler ultrasonic velocity sensors*

Doppler velocity sensors measure velocity by transmitting a continuous wave of sound at a known frequency into the flow. When that sound is reflected by a moving particle (suspended material) in the stream, the echo returns to the sensor at a different frequency. This Doppler Frequency Shift is directly related to the speed of the particle along the acoustic beam.

The Acoustic Doppler Current Profiler (ADCP) is described in Section 4.3.12.

The instruments to measure flow velocities are summarized in Table 4.4.

### 4.3.5 *Velocity measurements*

*Position in the cross-section*

The place of the verticals in a cross-section is strongly dependent on whether or not boats are used, anchoring facilities, width and cross-sectional shape of the channel, etc. Some examples:

1.  *Wading*

In small, shallow water courses, measurements are carried out by wading. The current-meter is supported on a graduated wading rod provided with a bottom plate. In most cases positioning is done in the cross-section, using a measuring tape, fastened onto both banks with pegs.

The position of the operator should not affect the flow conditions around the current-meter: the best position is diagonally downstream of the meter and an arm length from it. The rod should be kept vertical with the meter in the direction of flow.

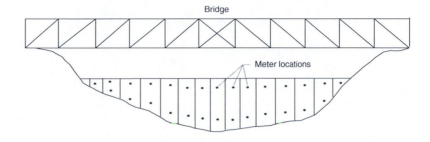

Figure 4.17. Measurement
from a bridge.

2.    *Measurement from a bridge (Figure 4.17)*
Advantages:
−   easy positioning
−   no vessel(s) required

Disadvantages:
−   sometimes hindrance from bridge piers and bridge abutments, which
    may affect the flow pattern. Piers and abutments must be cleaned of
    floating debris.

3.    *Measuring boat(s) along a cable across the channel (Figure 4.18)*
Advantages:
−   accurate and quick positioning; no anchoring problems.

Disadvantages:
−   blocking river traffic
−   only applicable for not too wide channels
−   facilities required to fix cable ashore

4.    *Anchored vessel(s)*
Advantages:
−   bigger ships can be used, offering more facilities
−   no limits to river width

Disadvantages:
−   anchorage equipment required
−   river depths to be sufficient
−   wind hindrance
−   much time required for shifting ship's position in measuring section

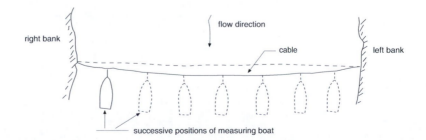

Figure 4.18. Measurement
along a cable.

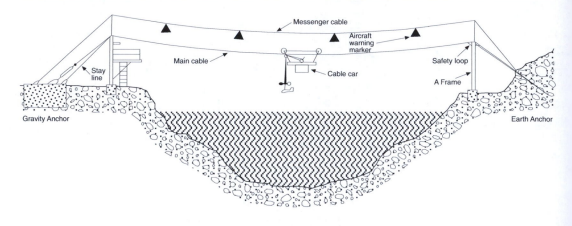

Figure 4.19. A typical manned cable way (after: Water Survey of Canada, 1984).

## 5.   Cable ways

Cable way systems are constructed to obtain measurements of streamflow or sediment transport in a cross-section. The use of cable ways eliminates the risks to operators of working from bridges and boats.

The position of the meter at each vertical and at any depth can be achieved by distance and depth counters, when using an unmanned cable way.

Figure 4.19 shows a manned cable way with cable car.

Cable ways are mainly used in steep rivers. The cable span usually varies from 40 to 400 metres. The main cable sag depends on the span, the cable diameter (19...29 mm) and the load.

*Number and location of verticals in the cross-section*
Number: see under Section 4.3.2 and the relevant ISO standard.
Location:
– normally, the verticals should be regularly spaced: equidistant
– in some cases (very irregular cross-section) spacing is to be adapted to the circumstances, as indicated in Figure 4.20
In judging the specific number of verticals (m) the following criteria – according to ISO standards – are to be applied.

Table 4.5. Number of verticals as a function of the channel width.

| Channel width (m) | Number of verticals m |
| --- | --- |
| b < 0.5 | 3 |
| 0.5 < b < 1.0 | 4 to 5 |
| 1.0 < b < 3.0 | 5 to 8 |
| 3.0 < b < 6.0 | 8 to 12 |
| b > 6.0 | 12 or more |

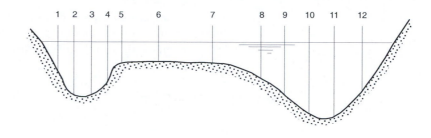

Figure 4.20. Location of verticals in an irregular cross-section.

*Number and location of points in the vertical*
Number: see under Section 4.3.2
Location: various approaches are possible and justified
– normally, the three point method as recommended in ISO standards will be applied, provided the velocity distribution is parabolic
– in some cases, when the velocity distribution is not parabolic or unknown, equidistant regular spacing is recommended
  If n points are used, then for instance the upper point as close as possible below the water surface, the lowest point as close as possible above the riverbed, and the other points evenly spaced between the upper and lowest points.
  Remark: the velocity measurement in the lowest point can be very inaccurate or – without being realized by the operator – even wrong due to the presence of obstacles, for instance high sand ripples, on the riverbed, as shown in Figure 4.21.

In judging the specific number of points (n) in a vertical, the following criteria can be applied.

### 4.3.6 Determination mean velocity in the vertical

Because of the relatively slowly changing conditions of the river flow in non-tidal rivers, the velocity measurements can be carried out by measuring the velocities in one vertical after the other along the cross-section.

Figure 4.21. Velocity measurement in an alluvial channel with sand ripples.

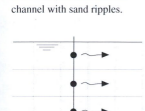

In order to measure the whole velocity profile in a reasonably short time, the velocity profile is sampled by a finite number of observations, provided the velocity distribution is parabolic.

Table 4.6. Number of points in the vertical as a function of the water depth.

| Water depth | Number of points n |
|---|---|
| d < 0.25 m | 1 |
| 0.25 < d < 0.50 | 2 |
| d > 0.50 | 3 or more |

The mean flow velocity in each vertical can be determined by any of the following methods depending on the available time and taking into account the accuracy, the width and depth of the water and the bed conditions.

1. *One point method*

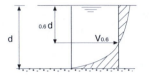

Figure 4.22. One point method.

$$\overline{v} = v_{0.6} \qquad (4.13)$$

where $v_{0.6}$ is the measured velocity at 0.6 d from the water surface.

2. *Two point method*

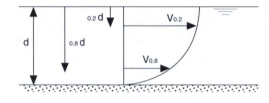

Figure 4.23. Two point method.

$$\overline{v} = 0.5(v_{0.2} + v_{0.8}) \qquad (4.14)$$

where $v_{0.2}$ and $v_{0.8}$ have been measured at 0.2 d and 0.8 d from the water surface.

3. *Three point method*

$$\overline{v} = 0.25v_{0.2} + 0.50v_{0.6} + 0.25v_{0.8} \qquad (4.15)$$

which is a combination of the one point method and the two point method.

4. *Five point method*

$$\overline{v} = 0.1(v_s + 3v_{0.2} + 2v_{0.6} + 3v_{0.8} + v_b) \qquad (4.16)$$

$v_s$ = velocity at surface
$v_b$ = velocity at bottom

5. *Graphical method (Figure 4.24)*
This method is used, if the velocity distribution is not parabolic or unknown.

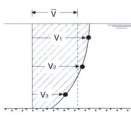

Figure 4.24. Graphical method.

Through the plotted results (measured velocities $v_1$, $v_2$, …etc) a vertical velocity distribution graph is drawn. Above the upper and below the lowest measuring point the line is to be extrapolated till water surface and bottom line respectively.

Using a planimeter the shaded area is determined. The average velocity $\bar{v}$ then is:

$$\bar{v} = \frac{\text{shaded area (in cm}^2) \times \text{hor. scale} \times \text{vert. scale}}{\text{total real water depth (m)}}$$

*Example:*

|  |  |
|---|---|
| shaded area | $4.98 \text{ cm}^2$ |
| water depth | $3.00$ m |
| horizontal scale | $1 \text{ cm} \approx 0.50 \text{ m/s}$ |
| vertical scale | $1 \text{ cm} \approx 1.00 \text{ m}$ |

$$\bar{v} = \frac{4.98 \times 0.50 \times 1.00}{3.00} = 0.83 \text{ m/s}$$

*6. Integration method*
This method is also used if there are doubts about the velocity distribution. The current-meter is lowered and raised through the entire vertical at a constant and very low speed (less than 5% of the mean flow velocity and less than 0.04 m/s). It is recommended to measure two cycles: one lowering and one raising the meter. The method is suitable for propeller current-meters and electromagnetic flow velocity sensors (not for cup-type meters). The mean velocity in the vertical – using a propeller current-meter – is obtained from the propeller's calibration curve and the average number of propeller revolutions (total number of revolutions divided by the total cycle time).

### 4.3.7 *Determination of the total discharge in the cross-section*

Three methods to calculate the total discharge are mentioned, each of them based on the measurement of the depth $d$ and the calculated depth average velocity $\bar{v}$ in a number of verticals.

*1. Graphical method (or depth velocity integration method)*
The discharge per unit width $q$ i.e. the product of the value of the mean velocity $\bar{v}$ at each vertical and the corresponding depth ($\bar{v} \cdot d$), should be plotted over the water surface line. A smooth curve may be drawn up connecting the $\bar{v} \cdot d$ points as shown in Figure 4.25.

The area enclosed by the unit discharge curve, $q = \bar{v} \cdot d$, is calculated using a planimeter, taking into account the scales of the graph.

*2. Mean Section method*
The cross-section is regarded as being made up of a number of panels or subsections, each bordered by two adjacent verticals. If $\bar{v}_1$ and $\bar{v}_2$ are the mean velocities at the first and second vertical respectively, and if $d_1$ and $d_2$ are the depths measured at the verticals I and II respectively

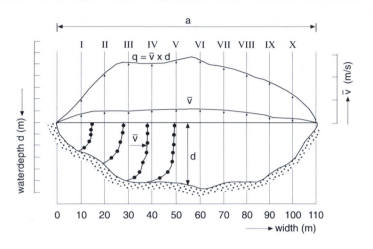

Figure 4.25. Depth velocity
graph (after: Hayes, 1978).

and '*b*' is the width between the said verticals, then the discharge of the
panel is to be calculated as

$$Q_p = \frac{\overline{v}_1 + \overline{v}_2}{2} * \frac{d_1 + d_2}{2} * b \qquad (4.17)$$

where $Q_p$ is the total partial discharge through the considered panel.
    This is to be repeated for each panel and the total discharge is the sum-
mation of the discharges per panel.

For the panels at the site (close to the bank) the same equation can be
used as above, whereas the velocity at the bank is taken as zero.
    One should realize, however, that the mean velocity in horizontal
direction towards the banks in many cases has a parabolic form and
therefore it may give a better estimate to calculate $Q_p$ for the panels near
the banks as

$$Q_p = \frac{2}{3}\overline{v}_1 * \frac{1}{2}(d_0 + d_1) * b \qquad (4.18)$$

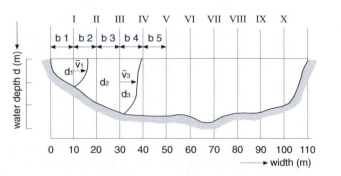

Figure 4.26. Mean Section
method (after: Hayes, 1978).

b  = the width from the bank to the vertical I
$\overline{v}_1$ = the mean velocity in vertical I
     The total discharge is the sum of all the calculated $Q_p$'s.

### 3.   *Mid Section method*

Assuming a straight line variation of $\overline{v} \cdot d$, the discharge in each section should be computed by multiplying $\overline{v} \cdot d$ by the corresponding width measured along the water surface line. This width should be taken to be the sum of half the width of the adjacent vertical to the vertical for which $\overline{v} \cdot d$ has been calculated, plus half the width of this vertical to the corresponding adjacent vertical on the other side. The discharge around vertical II is:

$$Q_p = \overline{v}_2 \cdot d_2 * \frac{1}{2}(b_2 + b_3) \tag{4.19}$$

in which
$b_2$ = the horizontal distance between vertical I and II
$b_3$ = the horizontal distance between vertical II and III

The value of $\overline{v} \cdot d$ in the two half-widths next to the banks should be taken as zero.
     The total discharge is a summation of all the calculated $Q_p$'s.

It is very important to use an appropriate measuring form. Table 4.7 is a discharge measuring form for the current-meter method, using a propeller current-meter and applying the three-point method.
– general information: location and time, measured water levels, propeller number and calibration, and the measuring time $t$
– field measurements: location of the verticals, measured water depths and measured propeller revolutions at three depths from the water surface
– computation of the total discharge

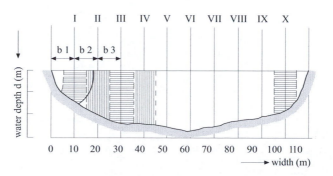

Figure 4.27. Mid Section method (after: Hayes, 1978).

| | | |
|---|---|---|
| Measured water levels: | River | : Lower Rhine |
| KM 895 MSL + 10.74 m | Location : | KM 897 |
| KM 910 MSL + 8.94 m | Date | : April 23, 1996 |
| | Time | : 09.00–12.00 hr |

| | |
|---|---|
| Propeller | Number of revolutions per sec. |
| $n < 0.63$    $v = 0.246\, n + 0.017$ (m/s) | $n = n_1/t$ |
| $n > 0.63$    $v = 0.260\, n + 0.008$ (m/s) | Measuring time $t = 50$ sec. |

## a)  *Field measurements*

| Sec-tion | Distance (m) | | Depth $d$ (m) | Propeller 0.2 $dv_{0.2\,d}$ | | Propeller 0.6 $d$ | | Propeller 0.8 $d$ | | Mean vel. $\overline{V}$(m/s) |
|---|---|---|---|---|---|---|---|---|---|---|
| | Ref.p. | LB | | Revol. $n_1$ | $v_{0.2\,d}$ (m/s) | Revol. $n_1$ | $v_{0.6\,d}$ (m/s) | Revol. $n_1$ | $v_{0.8\,d}$ (m/s) | |
| LB | 8.60 | 0 | 1.00 | | | | | | | |
| I | 18.60 | 10 | 2.24 | 103 | *0.544* | 99 | *0.523* | 85 | *0.450* | *0.51* |
| II | 28.60 | 20 | 3.16 | 141 | *0.741* | 134 | *0.705* | 115 | *0.606* | *0.69* |
| III | 38.60 | 30 | 3.72 | 153 | *0.804* | 149 | *0.783* | 129 | *0.679* | *0.76* |
| IV | 48.60 | 40 | 4.16 | 162 | *0.850* | 158 | 0.830 | 134 | 0.705 | *0.80* |
| V | 58.60 | 50 | 4.00 | 160 | *0.840* | 154 | 0.809 | 132 | *0.694* | *0.79* |
| VI | 68.60 | 60 | 3.08 | 138 | *0.726* | 132 | *0.694* | 114 | *0.601* | *0.68* |
| VII | 78.60 | 70 | 2.44 | 95 | *0.502* | 87 | *0.460* | 72 | *0.382* | *0.45* |
| RB | 90.20 | 81.60 | 1.62 | | | | | | | |
| | | | | | 3 dec. | | 3 dec. | | 3 dec. | 2 dec. |

Table 4.7a. A worked out discharge measuring form (measurements).

## b) Computation of total discharge by the *Mean section method*

| Section | Distance LB (m) | Depth $d$ (m) | $\overline{v}$ (m/s) | $b$ panel (m) | $d$ panel (m) | $\overline{v}$ panel (m/s) | $Q$ panel (m³/s) |
|---|---|---|---|---|---|---|---|
| LB | 0 | 1.00 | | | | | |
| | | | | *10* | *1.62* | *0.34* | *5.5* |
| I | 10 | 2.24 | *0.51* | | | | |
| | | | | *10* | *2.70* | *0.60* | *16.2* |
| II | 20 | 3.16 | *0.69* | | | | |
| | | | | *10* | *3.44* | *0.72* | *24.8* |
| III | 30 | 3.72 | *0.76* | | | | |
| | | | | *10* | *3.94* | *0.78* | *30.7* |
| IV | 40 | 4.16 | *0.80* | | | | |
| | | | | *10* | *4.08* | *0.80* | *32.6* |
| V | 50 | 4.00 | *0.79* | | | | |
| | | | | *10* | *3.54* | *0.74* | *26.2* |
| VI | 60 | 3.08 | *0.68* | | | | |
| | | | | *10* | *2.76* | *0.56* | *15.5* |
| VII | 70 | 2.44 | *0.45* | | | | |
| | | | | *11.60* | *2.03* | *0.30* | *7.1* |
| RB | 81.60 | 1.62 | | | | | |
| | | | | | | $Q$ total = | 158.6 |
| | | | | | 2 dec. | 2 dec. | 1 dec. |

c) Computation of total discharge by the *Mid section method*

| Section | Distance LB (m) | Depth d (m) | $\overline{v}$ (m/s) | b panel (m) | $b_{mids}$ (m) | $Q_{mids}$ (m³/s) |
|---------|------|------|------|------|------|------|
| LB  | 0     | 1.00 |      |      |       |       |
| I   | 10    | 2.24 | 0.51 | 10   | 10    | 11.4  |
| II  | 20    | 3.16 | 0.69 | 10   | 10    | 21.8  |
| III | 30    | 3.72 | 0.76 | 10   | 10    | 28.3  |
| IV  | 40    | 4.16 | 0.80 | 10   | 10    | 33.3  |
| V   | 50    | 4.00 | 0.79 | 10   | 10    | 31.6  |
| VI  | 60    | 3.08 | 0.68 | 10   | 10    | 20.9  |
| VII | 70    | 2.44 | 0.45 | 10   | 10.80 | 11.9  |
| RB  | 81.60 | 1.62 |      | 11.6 |       |       |
|     |       |      |      |      | Q total = | 159.2 |

|  | 2 dec. |  | 1 dec. |
|--|--------|--|--------|

Table 4.7b. A worked-out discharge measuring form (computations).

### 4.3.8  *Uncertainties in the velocity area method*

Usually, the degree of accuracy required will be based on a number of considerations, according to needs from research, design, construction, economy, management, etc. An important consideration is the extent to which the accuracy can be improved with reasonably increased effort.

*Composition of total error*

For the determination of a discharge, a number of components have to be measured. To optimize the measurements, it is necessary to know the accuracy that can be achieved when measuring each of the components. As the total error is composed of uncertainties in the measurement of the individual components, it is evident that, if one of the components is measured relatively inaccurate, this may affect the total uncertainty to such an extent that very accurate measurement of the remaining components becomes meaningless.

In general, a distinction can be made between errors of a systematic, and those of a stochastic (random) nature.

According to their origin, errors can be distinguished as being either due to the instrument used, or to measuring procedures and the processing of data. They can be systematic as well as stochastic.

When using velocity area methods, three components have to be measured, i.e. width, depth and flow velocity including flow direction. Each of these measurements will be subject to the uncertainties mentioned.

An examination of the accuracy of the instruments is not included (depends on the type and on the maintenance).

The systematic bias error of an instrument is related to the characteristic properties of the instrument.

The stochastic instrumental error, however, has to be included in the calculation of the total stochastic error. For this reason, standard deviations of stochastic instrumental errors, known from existing literature and research, are given below:

a) In ISO 748 for the measurement of distance, a relative error of 0.3% is indicated for a distance between 0 and 100 m, and 0.5% for a distance of 250 m. When the distance is measured electronically, an error as a percentage of the distance (for instance 0.5 to 1%), in addition to a fixed error of 0.5 to 2 m, has to be considered.

b) The instrumental error in the measured depth depends, to a large extent, on the composition of the river bed, which is critical if the sounding rod, lead or acoustic pulse of the echo sounder penetrates into the bed. An error of 1% is considered to be a reasonable approximation.

c) For the determination of flow velocity, two types of instruments are used: the cup-type and the screw-type (propeller) current-meter. The standard deviation of the stochastic calibration error of cup-type current-meters is less than 1%.

Investigations using a screw-type current-meter showed a relative standard deviation of 5% for a flow velocity of 0.2 m/s decreasing to 0.5% for a flow velocity of 2.5 m/s.

For electromagnetic sensors, an error of 1% is to be considered as a minimum value. In case of unstable zero values, the error will increase particularly for low flow velocities: for instance 10% for $v = 0.02$ m/s, and 3% for $v = 0.10$ m/s.

*Errors in mean flow velocity and depth*

In the investigation, special attention was paid to stochastic errors due to the methods used for determining the mean flow velocity in the cross-section and to the methods used for determining the depth in the section considered.

Apart from the instrumental error, the error in the mean flow velocity component can be considered as comprising three independent types of errors:

- error type I    $\sigma_f$ (measuring time), due to the restricted measuring time of the local point velocity in the vertical;
- error type II   $\sigma_s$ (number of points $n$ in the vertical), arising from the use of a restricted number of sampling points in the vertical. The calculated mean velocity in a vertical is therefore an approximation of the true mean velocity in that vertical;
- error type III  $\sigma_d$ and $\sigma_h$ (number of verticals $m$), of the same nature as error type II, due tot the restricted number of verticals

in the cross-section. The horizontal velocity profile $\sigma_h$ and the bed profile between two verticals $\sigma_d$ have to be determined by interpolation and therefore errors will be introduced.

*Error type I ($\sigma_f$)*
Taking a measuring time $t = 30$ seconds for high velocities or $t = 60$ seconds for low velocities, the error reaches an acceptable minimum value. The random fluctuation error $\sigma_f$ depends furthermore slightly on the number of points in the vertical.
Values of $\sigma_f$ are therefore combined with those of error type II (Table 4.8).

*Error type II ($\sigma_s$)*
Usually, the mean flow velocity in a vertical is calculated by the use of one of the existing computation methods. These methods result in an approximation of the true mean velocity at a certain moment.

Table 4.8 gives the error $\sigma_{f+s}$ for six well-known methods, as a combination of the error types I and II (see also Figure 4.6).

The following conclusions have been drawn by the relevant ISO standards:
a) The methods have a more general validity for larger rivers ($Q > 120$ m³/s) than for smaller rivers ($Q < 120$ m³/s).
b) The nature of the velocity profile in the vertical is sufficiently fixed by measurements at three points (method No. 4). The result can be improved by increasing the total measuring time, either by measurements at more than three points or by increasing time at each of the three points.

Table 4.8. The error $\sigma_{f+s}$ (combination of error types I and II) as a function of the number of points $n$ in the vertical.

| No. | Method or rule | Number of points $n$ | Standard deviation of the mean error $\sigma_{f+s}$ (%) (types I and II) |
|---|---|---|---|
| 1 | $\overline{v} = v_{0.6}$ | 1 | 8.2 |
| 2 | $\overline{v} = 0.96v_{0.5}$ | 1 | 6.5 |
| 3 | $\overline{v} = 0.5(v_{0.2} + v_{0.8})$ | 2 | 4.9 |
| 4 | $\overline{v} = 0.25v_{0.2} + 0.5v_{0.6} + v_{0.8}$ | 3 | 4.8 |
| 5 | $\overline{v} = \dfrac{1}{4}(v_{0.2} + v_{0.4} + v_{0.7} + v_{0.9})$ | 4 | 3.0 |
| 6 | $\overline{v} = 0.1v_{surf.} + 0.3v_{0.2} + 0.2v_{0.6} + 0.3v_{0.8} + 0.1v_{bed}$ | 5 | 2.7 |

Remarks:
1. $v_{0.6}$ is the velocity measured at 0.6 $d$ (depth related to water surface).
2. rule No. 4, the three point method, is commonly accepted as a practical and reliable method, provided the velocity distribution is parabolic.

Table 4.9. The error $\sigma_{d+h}$ as a function of the number of verticals $m$.

| Number of verticals $m$ | Relative standard deviation of error, % | |
| --- | --- | --- |
| | Criterion a: Verticals equidistant | Criterion b: Bed profile in the cross-section |
| 5 | 4.20* | 7.70 |
| 6 | 3.70* | 7.00 |
| 10 | 2.60 | 4.40 |
| 15 | 1.98 | 3.02 |
| 20 | 1.65 | 2.20 |
| 25 | 1.45 | 1.70 |

* extrapolated by author.

*Error type III ($\sigma_d$ and $\sigma_h$)*
Error type III is due to the approximation by interpolation of the bed profile ($\sigma_d$) and the horizontal velocity distribution between the verticals ($\sigma_h$).

In practice, both factors usually occur simultaneously. The measurement of flow velocity and depth takes place in a restricted number of verticals located in the cross-section. The selection of the number and location of the verticals is mainly based on personal judgment, taking into account the shape of the bed profile in the cross-section.

In general, it is known that the selection of too few verticals may lead to a considerable error.

In choosing the number and location of the verticals, the following criteria may be used:

a) Verticals equidistant

In this method, the number of verticals is decided beforehand and they are spaced equally across the width. In cross-sections where variations in profile and horizontal velocity distribution are gradual, equal discharge in the various sections is approximated.

b) Bed profile in the cross-section

Some surveyors will choose verticals according to irregularities of the profile read from an echogram, bearing in mind that the distances between verticals should not vary too much.

The results are given in Table 4.9 for both criteria described above. They show the standard deviation of error type III. (See also Figure 4.7).

*Basic equation*
The following simplified equation can be used for the calculation of the total stochastic error:

$$\sigma_1 = \sqrt{(\sigma_{d+h})^2 + \frac{1}{m}(\sigma_{f+s}^2 + \sigma_v^2 + \sigma_d^2 + \sigma_B^2)} \qquad (4.20)$$

where

$\sigma_1$    is the relative standard deviation of the total stochastic instrumental and sampling error;

$\sigma_{d+h}$    is the relative standard deviation due to the random sampling error of the depth profile and the velocity profile (error type III), as a function of the number $m$ of verticals.

$m$    is the number of verticals;

$\sigma_{f+s}$    is the relative standard deviation due to the random error (error types, I and II) as a function of the number of points $n$ in the vertical;

$\sigma_v$    is the relative standard deviation due to the random instrumental error in determining the velocity;

$\sigma_d$    is the relative standard deviation due to the random instrumental error in determining the depth;

$\sigma_B$    is the relative standard deviation due to the random instrumental error in determining the width.

*Example*

River    $B = 80$ m

$\overline{d} = 2.50$ m

$Q = 120$ m$^3$/s

The river discharge is measured in two different ways:

1. taking $m = 15$ verticals, 3 points in the vertical (rule 4)
2. taking $m = 5$ verticals, 1 point at 0.6 $d$ from the water surface (rule 1).

    In both cases the verticals were equidistant.

    The contributing stochastic errors are summarized in Table 4.10.

The total stochastic error is    $\sigma_1 = \sqrt{(\sigma_{d+h})^2 + \dfrac{1}{m}(\sigma_{f+s}^2 + \sigma_v^2 + \sigma_d^2 + \sigma_B^2)}$

Case 1    $\sigma_1 = 2.4\%$

Case 2    $\sigma_2 = 5.6\%$

Table 4.10. Stochastic errors contributing to $\sigma_1$.

| Nature of stochastic error | | Case 1 | Case 2 |
|---|---|---|---|
| $\sigma_v$ | Estimated | 1.0 | 1.0 |
| $\sigma_{f+s}$ | (Table 4.8) | 4.9 | 8.2 |
| $\sigma_d$ | Estimated | 1.0 | 1.0 |
| $\sigma_B$ | Estimated | 0.4 | 0.4 |
| $\sigma_{d+h}$ | (Table 4.9) | 1.98 | 4.2 |

Finally the total stochastic error $\sigma_{sto.}$ is combined with the total systematic error $\sigma_{sys.}$ as follows:

$$\sigma = \sqrt{\sigma_{sto.}^2 + \sigma_{sys.}^2}$$

Note:    For this section use was made of the International Standard ISO/TR 7178-1983: Investigation of the total error in measurement of flow by velocity area methods.

### 4.3.9  *Limited number of verticals (the $d^{3/2}$ method)*

The availability of observers, vessels and instruments may be the reason that in river cross-sections, velocities are only measured in a few verticals so that the total discharge $Q$ (m³/s) has to be calculated from the measurements in a limited number of verticals. See Section 4.3.8 about the influence of the number of velocity observations in the vertical and the number of verticals on the overall accuracy of the discharge.

Although it should not be applied indiscriminately, good use can be made of the formula of Chézy for a number of cases that only too few verticals have been measured to obtain a good estimation of the $q$-values curve for the other verticals.

The formula of Chézy is:

$$\overline{v} = C\sqrt{RS} \tag{4.21}$$

in which
$\overline{v}$ = average velocity in a vertical (m/s)
$C$ = roughness coefficient of Chézy (m¹ᐟ²/s)
$R$ = hydraulic radius, $R = A/P$, for wide rivers R $\approx \overline{d}$
$d$ = water depth (m)
$S$ = longitudinal slope of the water surface

If for some reason or another the number of measuring verticals is to be very limited and – therefore – no good idea can be formed about the velocity (thus $q$–) distribution in the cross-section, then sometimes a 'trick' is applied which enables a numerical calculation of the discharge. This is done by establishing a numerical relation between the velocities and depths in non-measured verticals. The procedure is as follows:

For every point in a cross-section the $q$-value (m³/s/m) in the vertical through that point can be written as

$$q = \overline{v}d$$

So for points I, II, III, etc. in a cross-section we find respectively

$$q_1 = \overline{v}_1 d_1 \quad q_2 = \overline{v}_2 d_2 \quad q_3 = \overline{v}_3 d_3$$

The 'trick' implies that, applying the Chézy formula, *it is assumed that C and S have the same value for the whole cross-section*, thus

$$(C\sqrt{S})_1 = (C\sqrt{S})_2 = (C\sqrt{S})_3 = \overline{C\sqrt{S}}$$

The discharges in the various points of a cross-section then relate to each other as:

$$q_1 : q_2 : q_3 = (C\sqrt{S})_1 d_1^{\frac{3}{2}} : (C\sqrt{S})_2 d_2^{\frac{3}{2}} : (C\sqrt{S})_3 d_3^{\frac{3}{2}}$$

or $q_1 : q_2 : q_3 = d_1^{\frac{3}{2}} : d_2^{\frac{3}{2}} : d_3^{\frac{3}{2}}$

In other words: the discharges per unit width are proportional with $d^{3/2}$. (When the Manning formula is used, $v :: d^{2/3}$ and thus $q :: d^{5/3}$, the method then becomes a $d^{5/3}$-method in stead of the $d^{3/2}$-method and the constant value $C\sqrt{S}$ becomes $\sqrt{S/n}$).

The procedure is now that $q$-values are calculated for many points of the cross-section, multiplying the $d^{3/2}$-values of these points by the $\overline{C\sqrt{S}}$-value as calculated from the v- and d-values of the measured sections I, II and III. Summing up all $q \times \Delta B$ values the total discharge is obtained. Graphically the discharge is obtained by planimetering the area between $q$-line and water surface line taking into account the scales of the graph. (See Figure 4.28).

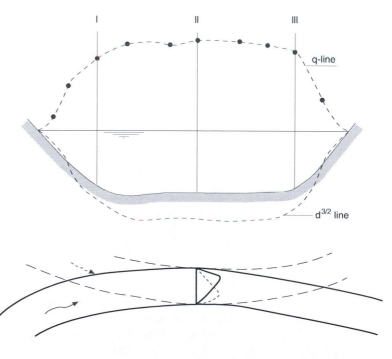

Figure 4.28. $d^{3/2}$-curve and discharge per unit width curve (after: Hayes, 1978).

Figure 4.29. Velocity distribution in a river bend.

Table 4.11 Calculation of $C\sqrt{S}$ values.

| Vertical | $d$ (m) | v (m/s) | $q$ (m²/s) | $C\sqrt{S}$ |
|---|---|---|---|---|
| I | 2.24 | 0.51 | 1.14 | 0.34 |
| II | 3.16 | 0.69 | 2.18 | 0.39 |
| III | 3.72 | 0.76 | 2.83 | 0.39 |
| IV | 4.16 | 0.80 | 3.33 | 0.39 |
| V | 4.00 | 0.79 | 3.16 | 0.40 |
| VI | 3.08 | 0.68 | 2.09 | 0.39 |
| VII | 2.44 | 0.45 | 1.10 | 0.29 |

*Example:*
Table 4.7 (Section 4.3.7) shows a complete discharge measurement, where water depths $d$ and flow velocities $\bar{v}$ have been measured in seven verticals. Now we apply the $d^{3/2}$ method by measuring the water depths in all verticals, but by measuring the flow velocities in only three verticals.

Table 4.11 shows the $C\sqrt{S}$ values for all the seven measured verticals.

The average value becomes $\overline{C\sqrt{S}} = 0.37$.

Application of the $d^{3/2}$ method for the verticals I, IV and VII leads to an underestimation of the total discharge. Taking the verticals II, IV and VI leads to an overestimation of the total discharge.

*Critical note with respect to the $d^{3/2}$-method*
The $d^{3/2}$-method is based on the *wrong* starting point that $C$- and $S$-values of the Chézy formula are characteristics of a cross-section. In reality $C$ and $S$ represent the average roughness and slope of a schematized longitudinal river reach.

The velocity distribution in a cross-section in first instance is determined by the curvature of upstream and downstream river parts whereas the cross-sectional profile depends mainly on whether the bottom is alluvial or not.

One should keep in mind the risk of errors when applying the $d^{3/2}$-method, because a well founded theoretical background is missing.

### 4.3.10 *The moving boat method*

In this section the moving boat method is linked with the use of one traditional current-meter (e.g. propeller) at one depth (instead of the use of more current-meters or the use of the ADCP).

Frequently on larger streams and in estuaries convential methods are impractical and involve costly procedures; this is particularly true during floods when facilities may be inundated or inaccessible, at remote sites

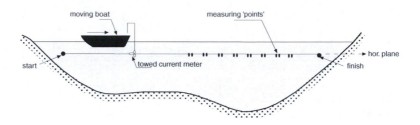

Figure 4.30. Principle of the moving boat method.

where no facility exist, or at locations where unsteady flow conditions require that measurements be made as rapidly as possible (tidal areas).

In other cases floating obstacles or river traffic require a flexible kind of discharge measurement that may not be interrupted due to this kind of unforseen events.

The moving boat technique is applicable to rapid measurement of rivers.

Restricting the number of measuring points in a vertical to only *one* and – at the same time – collecting as much information as possible on the horizontal velocity distribution is the principle of the so-called moving boat method. Therefore this method looks attractive, to speed up $Q$-measurements in wide rivers in the first instance.

With the moving boat method velocities are measured by suspending a continuously operating current-meter at a constant depth during an uninterrupted traverse of the boat across the stream in a prescribed line (section).

To determine the discharge, the following data are required:
– flow velocity in a great number of observation points;
– location of the observation points;
– water depth at the observation points.

The water depths are measured during the traverse with an echosounder. Positioning is done in principle for any point where velocity is measured. Positioning procedures may be chosen depending on the river width and the measuring method used (see Figures 4.31 and 4.32).

The recorded velocity represents the relative velocity of the water passing the current-meter. The flow velocity can be expressed by:

$$v_w = v_r \sin \alpha \tag{4.22}$$

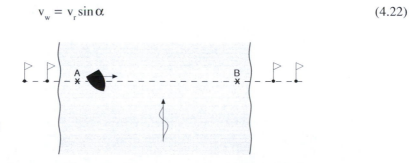

Figure 4.31. Positioning using range finder (after: Hayes, 1978).

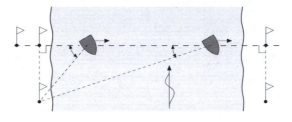

Figure 4.32. Positioning using sextant (after: Hayes, 1978).

or

$$v_w = \sqrt{v_r^2 - v_b^2} \qquad\qquad (4.23)$$

with

$v_w$ = flow velocity in measuring point
$v_r$ = velocity indicated by current-meter (= relative water velocity)
$\alpha$ = angle between the section and the orientation of the current-meter axis
$v_b$ = velocity of boat when sailing in the section, perpendicular to the stream direction

Different measuring methods can be applied.
1. for instance, if the boat is sailing exactly in a prescribed section line, during any period $\Delta t$ that the flow is measured ($v_r$), the position of the boat and the angle ($\alpha$) are observed. On the boat an angle indicator is installed for that purpose. The flow velocity is then expressed by $v_w = v_r \sin \alpha$
2. with another method the boat velocity $v_b$ at any measuring 'point' follows from consecutive position measurements. The boat is following the prescribed section line as well as possible, so that the flow direction is still about perpendicular to the boat direction. The water velocity follows from: $v_w = \sqrt{v_r^2 - v_b^2}$ .
3. measuring the boat-velocity $v_b$ and the angle $\alpha$ gives $v_w = v_b \cdot tn\,\alpha$.

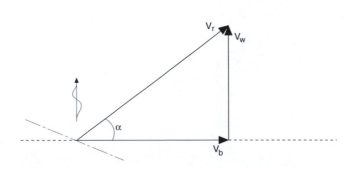

Figure 4.33. Diagram of velocity vectors.

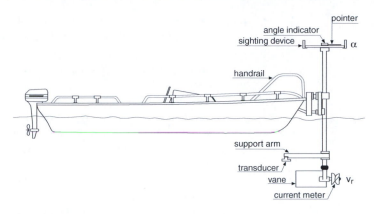

Figure 4.34. Layout of
equipment, moving boat,
(after: Hayes, 1978).

With this method simultaneous echo sounding is essential.

Applying this method, the current-meter need not to be visible from
the boat and, therefore, can be suspended by cable and at any depth
required.

During the traverse, velocity observations can be done at regular time
intervals (each observation for instance 20 or 30 s; counters to be 'frozen'
during that period). The measured $v_w$ is to be multiplied by a factor $k_v$ to
find the *mean velocity in the vertical* ($\overline{v}$).

$$\overline{v} = k_v \cdot v_w \tag{4.24}$$

When supporting the current-meter not too far below the water surface,
$k_v = 0.9$ can best be applied.

Final remarks:
- Application of the method is only recommended as long as the flow is
  clearly in one direction
- The moving boat method, having the advantage of quick and flexible
  procedures, can only be applied when very experienced surveyors and
  boat crews are available.
- Positioning can also be done using GPS equipment

### 4.3.11 *Float measurements*

Measurement of discharges using floats is done in extreme circumstances
when it is impossible to use a current-meter (excessive high velocities or
very low velocities) and in cases of reconnaissance.

For float measurements a measuring reach should be selected where
the flow is as uniform as possible: a straight channel with a uniform
cross-section.

Three sections are selected in this reach, at the beginning (1), mid-
way (2) and at the end (3) of the reach. The three sections are indicated

by six reference points, three on each bank, or all six on one bank. The sections must be spaced far enough to create a travel time of 20 seconds or more.

Various types of floats are used. Three well-known types are shown in Figure 4.35.

1. *Surface floats*. They have a depth of immersion less than 25% (preferable 10%) of the water depth. Examples: wooden blocks, coconuts, oranges or any kind of floating vegetation, sufficiently submerged to prevent wind effects. Surface floats may be used during floods.
2. *Double floats*. A small and low-resistant surface float is connected with a subsurface resistance body positioned at 0.6 of the water depth below the surface to directly obtain the mean velocity
3. *Rod floats*. They have a depth of immersion more than 75% (preferable 95%) of the water depth. However, they may not touch the bottom. In practice, the rod floats may be used only in artificial or other rectangular channels with a uniform cross-section, where the bed is free from obstacles and weeds.

The measuring procedure – as indicated in Figure 4.36 – is as follows:
- select roughly the number *m* of verticals between both banks in order to determine the discharge in Section 2.
- the first float is released far enough above Section 1 to adapt to the water flow velocity
- the float velocity is measured over the reach between the Sections 1 and 3 by measuring the distance *L* and the travel time *t* between both sections

$$v_{\text{float}} = L/t \tag{4.25}$$

It is assumed that this average velocity between the Sections 1 and 3 presents the float velocity in Section 2
- where the float is passing Section 2, the distance from the bank and the local water depth *d* are measured

Figure 4.35. Three types of floats (after: Jansen, 1979).

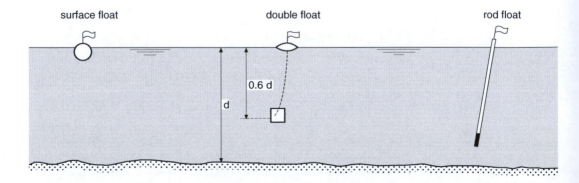

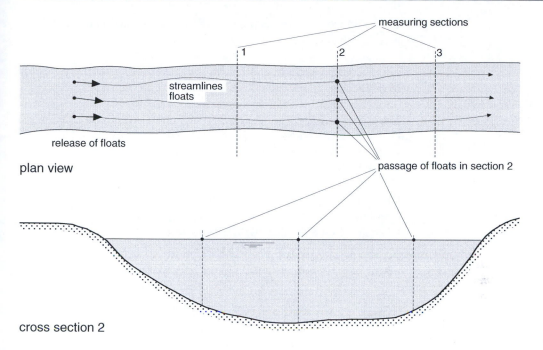

measuring sections

1   2   3

streamlines
floats

release of floats

plan view

passage of floats in section 2

cross section 2

Figure 4.36. Principle of float
measurements.

– the remaining floats are released one by one so that the float velocity $v_{float}$ and the water depth $d$ are measured in the desired number of $m$ verticals in Section 2

The various types of floats have reduction factors $k$ varying from 0.75 to 0.98. The mean flow velocity in the vertical reads:

$$\overline{v} = k \cdot v_{float} \tag{4.26}$$

The value of the reduction factor $k$ – also called the float velocity adjustment factor is given in Table 4.12 for surface floats and rod floats.
– surface floats: $k$ is a function of the channel roughness $n$
– rod floats: $k$ depends on the ratio of the immersed depth of float $y$ to the depth of water $d$

Table 4.12. Reduction factor $k$ for surface floats and rod floats.

| Surface floats | | Rod floats | |
|---|---|---|---|
| $n$ (m$^{-1/3}$ · s)$^*$ | $k$ (–) | $y/d$ | $k$ (–) |
| 0.029 – 0.037 | 0.78 | 0.10 | 0.86 |
| 0.021 – 0.028 | 0.84 | 0.25 | 0.88 |
| 0.017 – 0.022 | 0.87 | 0.50 | 0.90 |
| 0.014 – 0.019 | 0.89 | 0.75 | 0.94 |
| 0.012 – 0.016 | 0.90 | 0.95 | 0.98 |

* hydraulic radius 0.50 m < R < 2.50 m

Using surface floats in steep mountain rivers, a reduction factor $k = 0.75$ is applied by many river surveyors.

The discharge in Section 2 can be calculated using the mean section method or the mid section method.

The main sources of errors with float measurements are listed below
- too few verticals $m$
- non-uniformity in cross-sections 1, 2 and 3
- wind effects, especially with surface floats

Other methods of float measurements are:
- following floats with DGPS equipped vessels
- applying GPS floats, to be DGPS corrected off-line

### 4.3.12  *The Acoustic Doppler Current Profiler, ADCP*

Acoustic Doppler Current Profilers are relatively new instruments, used to measure discharges in open channels, according to the velocity-area method. Three different systems can be distinguished:
- downward looking sensors, vessel mounted. Used in the moving boat method, single determination of discharge. Can also be operated from a bridge (catamaran) or from a cableway. No restriction on the width of the river.
  This system is described in the standards ISO/TR 24578 and ISO/TS 24154.
- side looking /horizontal sensors, fixed to the side of the channel, continuous determination of discharges. Restrictions on the width of the river.
- upward looking sensors, bed mounted, continuous determination of discharges. They are normally used in smaller channels up to 5 m wide and 5 m deep. Sometimes bed mounted sensors are used in the acoustic (echo) correlation method (particle reflection in different scans).
  The side looking and the upward looking systems are described in ISO 15769.
  An example of a downward looking installation is the vessel mounted ADCP which is in use since the mid-1980's. A modern vessel-mounted ADCP is the so-called Broad Band ADCP, in production since 1992.

For the description of the Broad Band ADCP the following literature has been consulted:
1. Gordon R.L.
   Acoustic Doppler Current Profiler. Principles of Operation.
   R.D. Instruments, San Diego, Califomia, USA, January 1996.
2. Fieldservice Technical Paper 001. Broad Band ADCP. Advanced Principles of Operation. R.D. Instruments, San Diego, Califomia, USA, October 1996.
3. I.S.O. Technical Report, Measurement of liquid flow in open channels. Acoustic doppler current-meters.
   International Organization of Standardization (I.S.O.), Geneva 1998.

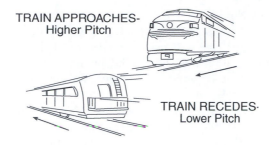

TRAIN APPROACHES-
Higher Pitch

TRAIN RECEDES-
Lower Pitch

Figure 4.37. Doppler Shift
when a train passes (after
RDI).

*The Doppler effect*

The *Doppler effect* is a change in the observed sound pitch that results
from relative motion. An example of the Doppler effect is the sound
made by a train as it passes (Figure 4.37). The whistle has a higher pitch
as the train approaches, and a lower pitch as it moves away from you.
This change in pitch is directly proportional to how fast the train is moving. Therefore, if you measure the pitch and how much it changes, you
can calculate the speed of the train.

When you listen to a train as it passes, you hear a change in pitch
caused by the Doppler Shift.

The Doppler principle can also be described using the water-wave
analogy:

Imagine you are next to some water, watching waves pass by you.
While standing still, you see eight waves pass in front of you in a given
interval. Now, if you start walking, toward the waves, more than eight
waves will pass by in the same interval. Thus, the wave frequency appears
to be higher. If you walk in the other direction, fewer than eight waves
pass by in this time interval, and the frequency appears lower. This is the
Doppler effect.

The Doppler shift is the difference between the frequency you hear when
you are standing still and what you hear when you move. If you are
standing still and you hear a frequency of 10 kHz, and then you start
moving towards the sound source and hear a frequency of 10.1 kHz, then
the Doppler shift is 0.1 kHz.

The equation for the Doppler shift in this situation is:

$$F_\mathrm{d} = F_\mathrm{s}(v/c) \tag{4.27}$$

where

$F_\mathrm{d}$   is the Doppler shift frequency
$F_\mathrm{s}$   is the frequency of the sound when everything is still
$v$   is the relative velocity between the sound source and the sound
     receiver (the speed at which you are walking towards the sound;
     m/s)
$c$   is the speed of sound (m/s). The sound transmitted by the ADCP is
     in the ultrasonic range between 300 and 3000 kHz (well above the
     range of the human ear)

Note that:
> If you walk faster, the Doppler shift increases.
> If you walk away from the sound, the Doppler shift is negative.
> If the frequency of the sound increases, the Doppler shift increases.
> If the speed of sound increases, the Doppler shift decreases.

*How ADCP's use backscattered sound to measure velocity*
ADCP's use the Doppler effect by transmitting sound at a fixed frequency and listening to echoes returning from sound scatterers in the water. These sound scatterers are small particles (suspended load) that reflect the sound back to the ADCP. Scatterers are everywhere in rivers. They float in the water and *on average they move at the same horizontal velocity as the water* (note that this is a key assumption!). Sound scatters in all directions from scatterers. Most of the sound goes forward, unaffected by the scatterers. The small amount that reflects back is Doppler shifted. When sound scatterers move away from the ADCP, the sound you hear is Doppler shifted to a lower frequency proportional to the relative velocity between the ADCP and the scatterer. The backscattered sound then appears to the ADCP as if the scatterers were the sound source. The ADCP hears the backscattered sound Doppler shifted a second time. Therefore, because the ADCP both transmits and receives sound, the Doppler shift is doubled, changing to

$$F_d = 2F_s(v/c) \tag{4.28}$$

Finally the Doppler shift only works when sound sources and receivers get closer to or further from one another, this is radial motion. Limiting the Doppler shift to the radial component, gives

$$F_d = 2F_s(v/c)\cos A \tag{4.29}$$

The ADCP measures only the velocity component parallel to the acoustic beams. A is the angle between the beam and the direction of flow.

*The Broad Band ADCP*
The narrow band ADCP was usable only in rivers and estuaries with depths greater than 3.5 m
The broad band ADCP was developed and tested in 1991, and is a feasible, rather accurate method for measuring discharges in tidally affected rivers and estuaries, as well as in rivers and canals with unsteady flow.

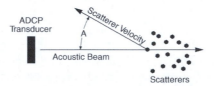

Figure 4.38. Relative velocity vector (after RDI).

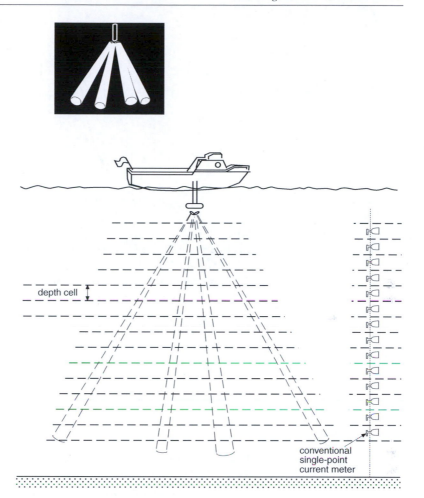

Figure 4.39. A typical Direct-Reading Broad Band ADCP (adapted from RDI).

*The broad band ADCP mounted at a moving boat*

The Acoustic Doppler Current Profiler ADCP is an accurate description of how the instrument determines flow velocities in a vertical profile:

1. Acoustic, as it uses sound waves to sense flow velocities;
2. Doppler, because it uses Doppler effect which is directly proportional to the flow velocity;
3. Current. The ADCP needs three beams to determine the three vector components of waterflow. The four-beam configuration provides two vertical vectors that the ADCP uses to check data integrity;
4. Profiler. The instrument measures flow velocities in a large number of points (depth cells) in the vertical by 'range-gating' the backscattered signal in time.

The ADCP makes a velocity profile for many depth cells, which can be from 5 cm in hight or more (depending on the frequency).

Figure 4.39 shows a four-beam configuration measuring in 14 cells.

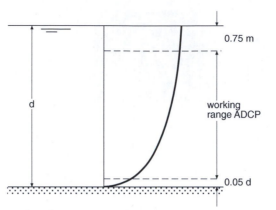

Figure 4.40. Working range ADCP in a vertical profile.

Each depth cell is comparable to a single current meter. Therefore an ADCP velocity profile is like a string of current meters uniformly spaced. Thus, we can make the following definitions by analogy:
– Depth cell size = distance between current meters
– Number of depth cells = number of current meters

There are two important differences between the string of current meters and an ADCP velocity profile. The first difference is that the depth cells in an ADCP profile are always uniformly spaced while current meters can be spaced at irregular intervals. The second is that the ADCP measures average velocity over the depth range of each depth cell while the current meter measures current only at one discrete point in space.

*Determination of mean velocities and total discharge*
The ADCP is used to measure flow velocities in a vertical profile. ADCP software is used to set measurement parameters, such as depth-cell size, maximum profiling depth, averaging method, etc.

The ADCP cannot be used to measure flow velocities near the water-surface (first 0.75 m) and the riverbed (0.05 $d$).

Users may extrapolate the measured velocity profile to the water surface and the riverbed using the Equation (4.7)

$$v_y/v_a = (y/a)^{1/n}$$

The mean velocity v in the vertical is calculated from the measured profile.

The ADCP is also used to measure discharges, according to the moving boat method. The motorized boat moves from the left bank to the right bank in a preselected cross-section. The measured velocities incorporate both the true flow velocity and the velocity of the moving boat. The latter is measured by the ADCP by bottom-tracking together with the measurement of water depths, where positioning is done by GPS equipment.

A single discharge measurement (moving boat), also called a transect, is a collection of ensembles (verticals). A typical ADCP transect will contain 2 to 3 times more ensembles than usually in the conventional method using propeller current meters (depending on the boat velocity).

The United States Geological Survey, USGS, carried out extensive investigation on ADCP's (see the References in this book).

### 4.4. SLOPE AREA METHOD

The slope area method is widely used to compute peak discharges after the passage of a flood. An ideal site is a reach of uniform channel in which the flood peak profile is defined on both banks by high water marks. From this information the water level slope, the cross-sectional area and the hydraulic radius are derived. Then the discharge is computed with the Manning formula or the Chézy formula.

The method can also be used to compute moderate discharges. In addition, the method can be applied in non-uniform channels.
The following items are discussed in this section:
– the conventional slope area method in a uniform channel
– the conventional slope area method in a non-uniform channel
– values of roughness coefficients for open channels
– the simplified slope-area method

*The conventional slope area method in a uniform channel*
In this method the discharge is computed – in most cases – with the Manning formula (4.5):

$$Q = \frac{1}{n} \cdot R^{\frac{2}{3}} \cdot S^{\frac{1}{2}} \cdot A$$

in which
$Q$ = discharge (m³/s)
$n$ = Manning's roughness factor (m$^{-1/3}$ · s)
$R$ = hydraulic radius $R = A/P$ (m)
$S$ = energy gradient (–)
  In a uniform channel the energy gradient is the same as the water level slope. Both are parallel to the stream bed (see Figure 4.41).
$A$ = cross-sectional area (m²)
$P$ = wetted perimeter cross-section (m)
The values of $A, P$ and $R$ remain approximately constant throughout the selected reach.

An alternative discharge equation is the Chézy formula:

$$Q = C \cdot R^{\frac{1}{2}} \cdot S^{\frac{1}{2}} \cdot A \tag{4.30}$$

in which

$C$ = Chézy coefficient written as $C = 18 \log \dfrac{12R}{k}$  (m$^{1/2}$/s)   (4.31)

$k$  = equivalent sand roughness of Nikuradse (m)

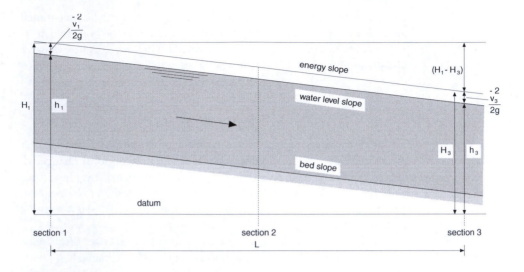

Figure 4.41. Slope area
method in a uniform channel.

The water level slope is derived from high water marks (peak discharges)
or calculated using water level gauges at both ends of the reach (moderate discharges).

$$S = \frac{(h_1 - h_3)}{L} \qquad (4.32)$$

The area of cross-section $A$, and its wetted perimeter $P$ are calculated from
measurements in at least three cross-sections: at the beginning $A_1$, $P_1$, mid-
way $A_2$, $P_2$ and at the end of the reach $A_3$, $P_3$. If the reach is substantially
uniform, then the mean area $\overline{A}$ and the mean wetted perimeter $\overline{P}$ are taken
as follows:

$$\overline{A} = \frac{(A_1 + 2A_2 + A_3)}{4} \qquad (4.33)$$

$$\overline{P} = \frac{(P_1 + 2P_2 + P_3)}{4} \qquad (4.34)$$

Application of the slope area method, which is much less accurate than
the other methods described, should only be considered as an *ad hoc*
method if the other methods are not feasible.

   The following criteria should be met:
1. The reach should have a uniform cross-section, be free from obstructions and from back water effects,

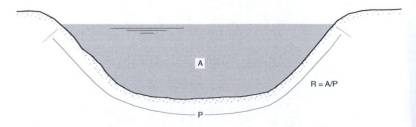

Figure 4.42. Definition of the
hydraulic radius *R*.

2. The length of the reach should be at least 75 times the mean depth, at least 5 times the mean width and preferable at least 300 metres,
3. The water surface fall in the reach should be greater than the velocity head and be at least 0.15 m,
4. The entire reach should have either subcritical flow or supercritical flow. The water level profile should not cross the critical depth line in a hydraulic drop or a hydraulic jump.

The procedure of the method is a follows:
- select a channel reach where the flow is approximately uniform, meeting the above mentioned criteria as well as possible,
- measure the cross-sectional profiles in the sections 1, 2 and 3.
  calculate the areas and wetted perimeters.
  calculate the mean values *A, P* and *R.*
- determine the water level slope *S* from high water marks, from water level gauges or by direct levelling.
- estimate the Manning or Chézy coefficient by comparing the actual roughness (bed material and vegetation) with a known roughness as indicated in literature (table 'Open Channel Hydraulics').
- calculate the flow velocity

$$\overline{v} = \frac{1}{n} \cdot R^{\frac{2}{3}} \cdot S^{\frac{1}{2}} \quad \text{or} \quad \overline{v} = C \cdot R^{\frac{1}{2}} \cdot S^{\frac{1}{2}} \quad \text{Is this credible?}$$

- calculate the discharge $Q = \overline{v} \cdot \overline{A}$

*Slope area method, an example*

*Basic equation* $\quad Q = \frac{1}{n} * R^{\frac{2}{3}} * S^{\frac{1}{2}} * A$

*Calculate discharge Q* with the following data:
- more or less uniform channel → uniform flow → $S = (WL1-WL3)/L$
- minor stream *B* < 25 m, full flow
  clean, winding, some pools, regular profile
- measured values:

*Computation*
- Table 4.14 → $n = 0.040$ m$^{-1/3}$s

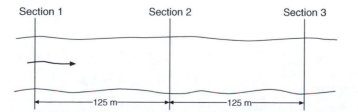

Figure 4.43. Example of the slope area method.

|  | Section 1 | Section 2 | Section 3 |
|---|---|---|---|
| water level WL (m) | MSL + 124.80 | | MSL + 124.10 |
| $A$ (m²) | 40.50 | 40.90 | 40.10 |
| $P$ (m) | 29.10 | 29.05 | 28.80 |

- $S = (124.80 - 124.10)/250 = 0.0028$
- $A = (40.50 + 40.90 + 40.90 + 40.10)4 = 40.60 \text{ m}^2$
- $P = (29.10 + 29.05 + 29.05 + 28.80)/4 = 29.00 \text{ m}$
- $R = A/P = 40.60/29.90 = 1.40 \text{ m}$
- $Q = 1.0/0.040 * (1.40)^{2/3} * (0.0028)^{1/2} * 40.60 = 67.2 \text{ m}^3/\text{s}$

In some natural streams – like rivers with floodplains – the channel roughness may differ over the entire cross-section. Then the cross-section is divided into parts with constant roughness, as indicated in Figure 4.44.

The discharge becomes $Q = S^{\frac{1}{2}} \cdot \sum_{i}^{n} C_i \cdot R_i^{\frac{1}{2}} \cdot A_i$

*The conventional slope area method in a non-uniform channel*
Uniform reaches can rarely be found in natural streams.

Therefore the Equations 4.5 and 4.30 are also used for non-uniform channels.

Equation 4.32 needs to be modified as follows:

$$S = \frac{(H_1 - H_3)}{L} \quad \text{with } H_1 = h_1 + \frac{v_1^2}{2g} \quad \text{and } H_3 = h_3 + \frac{v_3^2}{2g}$$

$$S = \frac{\left( h_1 - h_3 + \frac{v_1^2}{2g} - \frac{v_3^2}{2g} \right)}{L} \tag{4.35}$$

Figure 4.45 shows a non-uniform, slightly contracting reach.

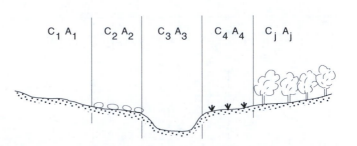

Figure 4.44. Cross-section divided into parts with constant roughness.

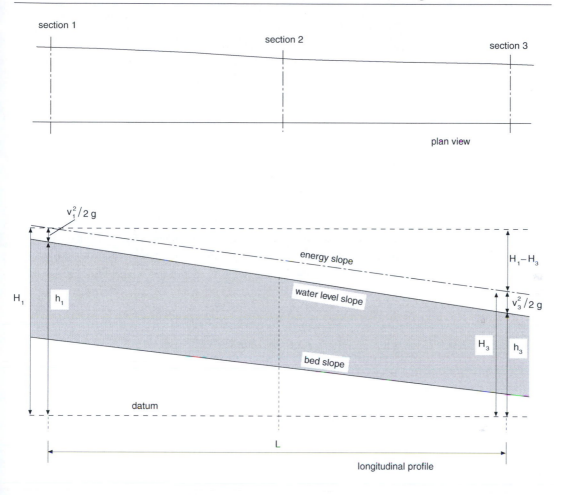

Figure 4.45. Slope area method in a non-uniform channel.

In the case of non-uniform reaches, the reach should preferable be contracting (converging) rather than expanding (diverging). Expanding reaches have diverging flow patterns, which may lead to unknown eddy losses.

*Values of roughness coefficients for open channels*
The slope area method is a very approximate way of discharge calculation, mainly caused by the estimation of the channel roughness coefficient, which is very arbitrary.

Various handbooks and the standard ISO 1070 provide Manning and Chézy coefficients.

Table 4.13 gives Manning's coefficient $n$ for channels with relatively course bed material and not characterized by bed forms, like ripples, sand dunes, etc.

In most natural channels the roughness coefficient is a function of:
-   bed roughness including bed forms and bank irregularity
-   effect of vegetation (season dependent)
-   water depth and channel slope

Table 4.13. Manning's coefficient $n$ as a function of bed material size.

| Bed material | Size of bed material (mm) | Manning's coefficient $n$ (m$^{-1/3}$ · s) |
|---|---|---|
| Medium gravel | 6–20 | 0.0195–0.022 |
| Course gravel | 20–60 | 0.022–0.027 |
| Cobbles | 60–200 | 0.027–0.034 |

Table 4.14 gives a selection of Manning's coefficients for those channels (taken from ISO 1070 and Ven Te Chow, 1959).

Visual estimating of $n$ values for mountain rivers during on site inspection is generally difficult. Some river engineers (Jarret, 1984) developed an equation to estimate Manning's $n$ for mountain rivers with cobble – or boulder – bed material as a function of slope and hydraulic radius:

$$n = 0.32S^{0.38} \cdot R^{-0.16} \qquad (4.36)$$

This empirical equation is defined for:
– slopes $0.002 < S < 0.030$
– hydraulic radius $0.50 < R < 2.00$ m
  The standard error for Equation 4.36 is 25–30%

Table 4.14. Manning's coefficient $n$ as a function of profile irregularity and vegetation.

| Type of channel and description | Manning's coefficient $n$ (m$^{-1/3}$ · s) Minimum | Normal | Maximum |
|---|---|---|---|
| *Excavated or dredged* | | | |
| a) Earth, straight and uniform | | | |
| 1. Clean, recently completed | 0.016 | 0.018 | 0.020 |
| 2. Clean, after weathering | 0.018 | 0.022 | 0.025 |
| 3. With short grass, some weeds | 0.022 | 0.027 | 0.033 |
| b) Channels not maintained | | | |
| 1. Clean bottom, brush on sides | 0.040 | 0.050 | 0.080 |
| 2. Dense brush, high stage | 0.080 | 0.100 | 0.140 |
| *Natural streams* | | | |
| a) Minor streams, on plain, $B < 30$ m | | | |
| 1. Clean, straight, regular profile. | 0.025 | 0.030 | 0.033 |
| 2. Clean, winding, some pools and shoals | 0.033 | 0.040 | 0.045 |
| 3. Same as 2, but irregular profile | 0.045 | 0.050 | 0.060 |
| b) Mountain rivers, at high stages | | | |
| 1. Gravel bottom with few boulders | 0.030 | 0.040 | 0.050 |
| 2. Cobbles with large boulders | 0.040 | 0.050 | 0.070 |
| c) Flood plains | | | |
| 1. Pasture, cultivated areas | 0.025 | 0.035 | 0.050 |
| 2. Brush and trees | 0.035 | 0.075 | 0.160 |

The Chézy coefficient is calculated from the Manning's coefficient as follows:

$$C = \frac{R^{\frac{1}{6}}}{n} \qquad (4.37)$$

*The simplified slope area method*
The reliability of computed discharges, using the conventional slope area method, depends largely on the estimation of the roughness coefficient.

Riggs developed a simplified slope area method for estimating flood discharges in natural channels (Riggs 1976), in which discharge is related to the cross-sectional area $A$ and to the water surface slope $S$.
– conventional method    $Q = f(n, R, S, A)$
– simplified method       $Q = f(S, A)$
Two assumptions are made:
1. roughness $n$ and slope $S$ are closely related
2. hydraulic radius $R$ and cross-sectional area $A$ are closely related
   The empirical equation is as follows:

$$\log Q = 0.191 + 1.33 \log A + 0.05 \log S - 0.056 (\log S)^2 \qquad (4.38)$$

Where $Q$ is in m³/s and $A$ is in m².
The standard error for this equation is about 20%.

The worlds maximum observed floods – according to the IAHS catalogue – can be expressed with $Q = 500\, A_d^{0.43}$, where $Q$ is the discharge in *m³/s* and $A_d$ is the drainage area in km² $(A_d \geq 100\ \text{km}^2)$. For most European rivers the highest floods are much lower.

## 4.5  DILUTION METHODS

In channels where cross-sectional profiles are difficult to determine (for instance in steep turbulent mountain streams) or where the flow velocities are too high to be measured with the normally used current-meters, the so-called dilution methods (or tracer methods) can be applied.

The dilution methods are based on the continuity principle; quantities of water and tracer material through one section have to pass through other sections as well. Permanent flow is assumed during the measurement.

The discharge is calculated from the degree of dilution – the dilution ratio – by the flowing water of an added tracer solution.

Dilution methods of measuring discharge have been known since at least 1863.

Mainly four different methods have been described in literature (see also the relevant ISO standards):
1. Constant rate injection method
2. Simplified constant rate injection method

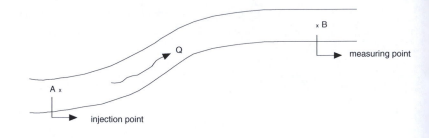

Figure 4.46. Constant rate injection method.

3. Sudden injection (integration) method
4. The cloud velocity method

Only the methods 1, 2 and 3 are discussed in this course.

An advantage of the dilution method is that no measurements of the geometry of the cross-section are required.

*Constant rate injection method*
Starting at $t = 0$ and during a certain time period a constant amount ($q$) of tracer solution with a concentration $C_1$ is added to the river at the injection point A. In a point $B$, being far enough downstream of $A$ to expect complete mixing of river water and tracer material, the river is continuously sampled and the tracer concentration is measured. After some time the tracer concentration $C_2$ in $B$ remains constant and the measurement is finished.

The following equation can be set up:

$$Q \cdot C_0 + q \cdot C_1 = (Q + q) \cdot C_2 \qquad (4.39)$$

where
$Q$ = river discharge (m³/s)
$q$ = injection rate (m³/s)
$C_0$ = natural tracer concentration of the river

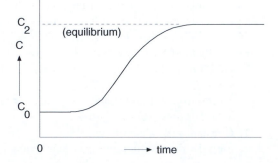

Figure 4.47. Mixing as a function of time.

$C_1$ = tracer concentration of injected solution
$C_2$ = concentration after complete mixing

By measuring the concentrations $C_0$, $C_1$ and $C_2$ and the rate $q$, the discharge of the river can be computed.

$$Q = q\frac{C_1 - C_2}{C_2 - C_0}$$    (4.40)

The ratio $N = \dfrac{C_1 - C_2}{C_2 - C_0}$ is called the dilution ratio.

If $C_2$ is small compared with $C_1$, then $Q = q\dfrac{C_1}{C_2 - C_0}$.

If table salt (NaC1) is used as a tracer, the concentration is generally calculated from conductivity measurements.

Tracer injection is usually done in the centre line of the river using the following equipment:
– the Mariotte bottle, with a volume $V = 25$ liters, for discharges $Q < 0.5$ m$^3$/s
– constant level tanks or volumetric pumps for discharges $Q > 0.5$ m$^3$/s.

*The Mariotte bottle*
The constant amount $q$ of tracer solution is added to the channel, using the bottle of Mariotte. This is a plastic vessel provided with a small valve close to the vessel's bottom. The opening on top of the vessel can be closed with a rubber stopper. The stopper has been perforated by a plastic pipe which is open at both ends. The length of the pipe is such that its bottom end is situated a few centimetres above the valve.
   The Mariotte bottle is operated as follows:
– fill the vessel with tracer solution (20 to 25 liters) and place it in the desired position at the injection point;
– open the valve. The water level in the plastic pipe will drop immediately to its bottom end, causing at the same time a low underpressure of the air in the upper part of the vessel (the stopper shall be well tapped);
– from this moment the pressure at the bottom end of the plastic pipe is the atmospheric pressure. The outflow $q$ through the valve is constant because the head $h_1$ over the valve remains constant (the head equals the small vertical distance between bottom end plastic pipe and centre line valve) until the water level in the vessel reaches the bottom end of the pipe. While the tracer solution is leaving through the valve, air is entering into the vessel through the pipe;

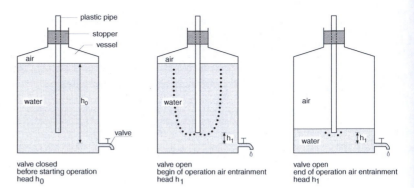

Figure 4.48. The operation of
the Mariotte bottle.

– a few seconds after the outflow has been started, its discharge can be
  measured volumetrically using a measuring glass and stopwatch.
  It is recommended to repeat this measurement at the end of the
  operation.
The total time of injection $t$ depends on the vessel's volume $V$ and the
outflow capacity of the valve. This capacity $q$ is governed by the head $h_1$
over the valve and its opening. The maximum injection period is $t = V/q$.
The outflow capacity can be changed by changing the valve's opening or
by changing the height of the pipe's bottom end.
    For most Mariotte bottles the outflow can be adjusted as $15 < q < 50$
($10^{-6}$ m$^3$/s).

The volume $V$ of the Mariotte bottle or the constant level tank
depends on the desired injection rate $q$ (derived from the desired dilution ratio $1000 < N < 30.000$) and the desired injection time, $5 < t <$
30 minutes.

*Tracer material*
Two groups are to be identified:
1. Colouring matter, like e.g. uranin (biological material) or rhodamine.
   The concentrations are measured with fluorometers;
2. Salt, like the common NaC1 (natural material). Concentration can be
   measured by measuring conductivity and temperature, or by titration;
   The conductivity, usually expressed in μS/cm depends strongly on the
   water temperature. Conductivity meters should be very sensitive and
   compensate for this effect.
General requirements for tracer material are:
– easy to dissolve;
– no absorption of tracer material by suspended or bottom sediments or
  by vegetation in the channel cross-section;
– no environmental pollution (easy to degrade);
– low costs.
The cheapest tracer is sodium chloride (NaC1), which is the ordinary
table salt. The required quantity is not particularly large. The solubility is
350 kg/m$^3$ at a water temperature of 15°C.

*Mixing length*

The mixing length is defined as the distance between the injection point and the sampling (measuring) point, in which the solution has been mixed completely over the cross-section.

This length depends on the mixing capacity which is a function of the turbulence.

The mixing length can be shortened if the tracer is evenly spread over the full width of the channel in the injection section.

When the constant rate injection method is used, then the mixing length is complete (100%) if the tracer concentrations are equal at all points in the cross-section. Complete mixing never occurs. In a practical sense 98% mixing is agreed as an adequate degree of mixing.

The standard ISO/TR 11656 gives a large number of empirical and theoretical relations to calculate the mixing length X for a degree of mixing 98%.

The Rimmar relation is a theoretical one. The Day formula is an empirical one.

Using a point injection in a small river, the minimum length required for complete mixing is given by the formula of Rimmar:

$$X_{min} = \frac{0.13B_s^2 \cdot C \cdot (0.7C + 2\sqrt{g})}{g \cdot d} \tag{4.41}$$

where

$B_s$ = average width at the water surface in the measuring reach (m)
$d$ = average depth in the same reach (m)
$C$ = Chézy coefficient for the reach, and $15 < C < 50$ $(m^{1/2}/s)$
$g$ = acceleration due to gravity $(m/s^2)$

Table 4.15 shows the mixing lengths for a small brook:
Width $B_s$ = 1.50 m, slope S = 0.003 and Mannings roughness $n$ = 0,030 $(m^{-1/3} \cdot s)$.

Table 4.15. Mixing lengths (Rimmar) for a small brook.

| Waterdepth $d$ (m) | Theoretical discharge $Q$ (m³/s) | Mixing length $X$ (m) |
|---|---|---|
| 0.10 | 0.041 | 89 |
| 0.15 | 0.077 | 66 |
| 0.20 | 0.120 | 53 |

Some authors, like Day, 1977, show that mixing lengths in mountain streams can be quickly estimated from a single geometric parameter, the mean flow width, using a simple equation:

$$X_{min} = 25B \tag{4.42}$$

*Simplified constant rate injection method (imitation method)*
In the simplified method the dilution process in the river is imitated in a
bucket:

– in the river the discharge $Q$ will dilute the Mariotte outflow $q$ which
  has a salt concentration $C_1$ and a conductivity $EC_1$.
  After complete mixing the maximum conductivity in the sampling
  section becomes $EC_2$.
– in the bucket a very small amount of the salt solution $V_s$ (between
  1 and 10 ml) is pipetted. Then an amount of river water $V$ is added to
  the salt solution until the mixture becomes the conductivity $EC_2$. The
  ration $N = V/Vs$ is the dilution ratio.
  The discharge of the river is calculated from the formula:

$$Q = \left(\frac{V}{V_s}\right) \cdot q \qquad\qquad (4.43)$$

*Apparatus*
The complete equipment, necessary to carry out the simplified constant
rate injection method is as follows:

– 1 Mariotte bottle (with frame to install the vessel above the water
  level)
– 1 funnel (to fill the vessel with water and salt)
– 2 plastic tanks (buckets 10 to 12 liters)
– 1 stick (to stir the water salt mixture)
– 2 measuring cylinders (500 ml)
– 2 pipettes (2 ml and 5 ml)
– 1 clock (stopwatch)
– 1 electro conductivity meter (range 0 to 5000 µS/cm)
– colouring matter Uranin (in order to find complete mixing over the
  cross-section, defining the mixture length)
– salt (about 4 to 8 kg per measurement with a vessel of 25 liters)

*The procedure*
1. Select the location following the criteria:
   – turbulent flow
   – no tributaries
   – no absorption of tracer material by sediments or vegetation
   – preferably normal flow from bank to bank
2. Inject Uranin solution in the injection point, in order to find the location of the sampling section.
3. Clean the Mariotte bottle, funnel, plastic buckets, stick, measuring
   cylinders and pipettes.
4. Prepare the salt solution in one of the buckets by mixing thoroughly
   4 to 8 kg salt with 25 liters river water, and keep it in the Mariotte bottle. Take care that no undissolved salt is in the bottle.
5. Pour river water into the second bucket and measure the conductivity.
   $EC_0$ (normally between 100 and 500 µS/cm).

Table 4.16. Measuring form simplified constant rate injection method.

*SALT DILUTION METHOD* Measuring Form

| River | : Silver Creek | Instrument | : Conductivity meter |
|---|---|---|---|
| Date | : April 7, 2007 | Mark | : |
| Time | : 11.00 h | Type | : |
| Observer(s) | : Waterman | | |

| | | |
|---|---|---|
| $EC_0$ | Of the river water before salt injection | 441 $\mu$S.cm$^{-1}$ |
| $q$ | Outflow from the Mariotte bottle | $45.5.10^{-3}$ $\ell$/sec |
| $EC_2$ | Maximum *EC* of the river water after the salt injection | 824 $\mu$S.cm$^{-1}$ |
| $V_s$ | Pipetted amount of salt solution | $5.10^{-3}$ litres |
| $V$ | Amount of water to be added to the pipetted salt solution | 3.025 litres |
| $Q$ | Discharge $Q = \dfrac{V}{V_s} \cdot q$ | $\dfrac{3.025}{5} \times 45.5 = 27.5$ $\ell$/sec |

TABLE 1

| time(s) | EC($\mu$S/cm) | | |
|---|---|---|---|
| 0 | 441 | 300 | 812 |
| 15 | 445 | 315 | 816 |
| 30 | 463 | 330 | 819 |
| 45 | 488 | 345 | 823 |
| 60 | 515 | 360 | 825 |
| 75 | 540 | 375 | 824 |
| 90 | 564 | 390 | 824 |
| 105 | 592 | 405 | 824 |
| 120 | 621 | 420 | 824 |
| 135 | 653 | 435 | 819 |
| 150 | 687 | 450 | 805 |
| 165 | 711 | 465 | 791 |
| 180 | 736 | 480 | 775 |
| 195 | 756 | | |
| 210 | 777 | | |
| 225 | 780 | | |
| 240 | 783 | | |
| 255 | 791 | | |
| 270 | 800 | | |
| 285 | 806 | | |

$EC_2 = 824$ $\mu$S.cm$^{-1}$

TABLE 2

| V(litres) | EC($\mu$S/cm) | |
|---|---|---|
| 0 | | |
| 0.5 | 2630 | |
| 1.0 | 1567 | |
| 1.5 | 1203 | |
| 2.0 | 1016 | |
| 2.5 | 906 | |
| 2.6 | 889 | |
| 3.025 | 824 | V = 3.025 litres |

the amounts *V* in this
table are cumulative
litres clean river water
(procedure step 11)

6. Install the Mariotte bottle on the right place where the salt solution is injected into the river (injection point).
7. Open the valve and measure after a few seconds the outflow, using a measuring cylinder and stopwatch. Repeat this measurement and take the average value of the outflow $q$. Keep the cylinder with salt solution apart (see item 10).
The valve is now permanently open (begin of operation).

8.  Walk to the sampling section with stopwatch and conductivity meter, and measure the conductivity every 15 seconds. Record the measurements into table 1 of the measuring form.
9.  Stop the measurement (and the injection) when the electric conductivity in the sampling section remains constant. This value will be the maximum conductivity $EC_2$.
10. Take an exact amount $V_s$ with the pipette ($1 < V_s < 5$ cc) of the salt solution which has been kept apart in step 7, and put it in a clean bucket.
11. Pour the clean river water into this bucket – using one of the measuring cylinders – in order to mix with the small amount $V_s$ of salt solution until the conductivity of this mixture equals the maximum conductivity $EC_2$ of the river water, measured under 9. The total amount of clean river water is $V$ (use table 2 of the measuring form).
12. Calculate the river discharge with ($Q = q \cdot V/V_s$)

*Example of the dilution measurement*
Table 4.16 shows a measuring form to measure the river discharge with the simplified constant rate injection method. In this example the dilution rate has been $V/V_s = 3025/5 = 605$.

A practical remark: before the process is imitated in the bucket, the amount $V_s$ of pipetted salt solution shall selected carefully.
–  a very small $V_s$ may lead to inaccuracy
–  if it is too much, then the content $V$ of the bucket may be too small
A correct amount $V_s$ can be estimated using the expression of the dilution ratio $N = Q/q = V/V_s$, hence $V_s = qV/Q$.
–  the Marriotte outflow $q$ has been measured volumetrically
–  take $V$ at about 50% of the capacity of the bucket
–  estimate roughly a value for $Q$ (mean width $\times$ mean depth $\times$ mean flow velocity)
–  calculate $V_s = q . V / Q$

## 4.6  STAGE DISCHARGE METHOD

### 4.6.1  *Introduction*

A continuous record of discharges in a gauging station may be computed from a continuous record of stages using the stage discharge relation, also called the rating curve.

Figure 4.49 shows a stage discharge relation for a station along a river.

The stage discharge relation is defined by plotting on graph paper the measured discharge as the abscissa and the corresponding stage as the ordinate. The shape of the curve is a function of the geometry of the river reach downstream of the station.

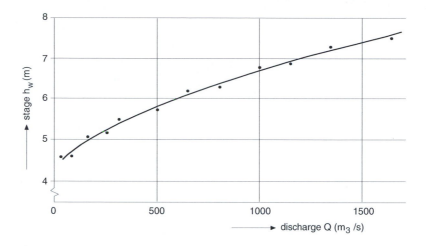

Figure 4.49. Typical stage discharge relation.

The plotted measurements have been collected using one of the single measurement methods; in most cases the velocity area method.

It is recommended that measurements be available throughout the whole range, as shown in Figure 4.49.

When plotted using a linear scale of both axi's, the relation is concave downwards, showing the form of a parabola. When plotted on log-log paper one or more straight lines are presented.

At many stations the discharge is a unique function of stage: the unique rating curve, not affected by downstream disturbances. At other stations the relations, especially at flood stages and low flows, may be distorted or they are unstable: non-unique rating curves.

Many measurements are necessary at a new gauging station, to define the stage discharge relation. Once the rating curve has been established, it is recommended to follow possible changes by periodic measurements; at least ten per year.

Assuming that the discharge measurements are properly made and that they apply to a stable stage discharge relation, it is convenient to derive a mathematical equation $Q = f(h_w)$.

In this section the following items are discussed:
– controls
– determination of the unique rating curve
– extrapolation of rating curves
– non-unique rating curves

### 4.6.2  Controls

The relationship between discharge and stage is generally set up for a pre-selected cross-section of a river or an artificial channel. This is particularly important for the stage discharge method and the slope stage discharge method. The pre-selected cross-section is also referred to as a control. A control is governed by the geometry of the cross-section and usually by the physical features of the river downstream of the section.

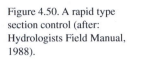

Figure 4.50. A rapid type section control (after: Hydrologists Field Manual, 1988).

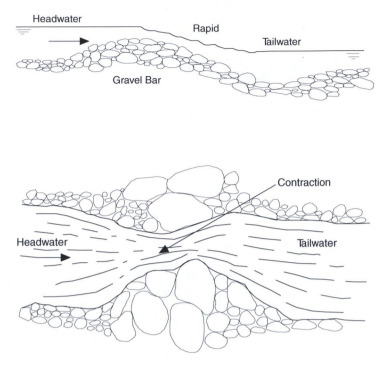

Figure 4.51. A contraction type section control (after: Hydrologists Field Manual, 1988).

The characteristics of the control determine the behaviour of the stage discharge relationship (shape, unique or non-unique, etc.).

Controls are classified as follows:

*section control*: natural or artificial local narrowing of the cross-section (waterfall, rock bar, gravel bar) creating a zone of acceleration. Disturbance downstream the control will not be able to pass the control in upstream direction.

*channel control*: a cross-section in which no acceleration of flow occurs or where the acceleration is not sufficient to prevent passage of disturbances in upstream direction. The rating curve depends on the geometry and the roughness of the river downstream of the control. The length of the downstream river – affecting the rating curve – depends on the discharge and on the energy slope. $L :: \dfrac{Q}{S}$.

*structure control*: artificial local narrowing of the cross-section like weirs and flumes discharging under free flow conditions.

Figures 4.50 and 4.51 are examples of section controls.

In many situations the cross-section is a complex control: during low discharges the section is a section control or a structure control, during peak flow the section becomes a channel control.

Once the control is used for measurement of water level and discharges, it is called a flow gauging station. Usually – at least in flat rivers – the gauging station is a channel control. However, if possible, a section control with a permanent narrowing should be selected.

### 4.6.3 *Determination of the unique rating curve*

When a number of discharges in one cross-section have been measured, it is possible to determine the relationship between the discharge of the section and the corresponding water level: the stage discharge curve. The most accurate curve is obtained when the measurements are evenly spread over the complete range of the occurring water levels. However, in general the observed discharges will not be situated on a smooth curve.

Regarding the shape of the stage discharge curve, the following remark must still be made. When the measurements are plotted on logarithmic paper, the stage discharge relation for a river with a more or less uniform longitudinal profile will be a straight line for the lowest water stages. In higher water stages the relation will often be smoothly curved because of the flooding of storage areas: flood plains, lakes, etc.

The results of discharge measurements and the corresponding stages are usually indicated as points in a $Q$-$h_w$ diagram, in which discharges are plotted as abscissae and stages as ordinates. Due to varying natural conditions during a series of measurements, it is obvious that the plotted points will not be located on one smooth curve. Therefore, a regression curve has to be drawn which fits the points as good as possible.

Various methods are used to determine the most probable regression curve. Although smoothness of the curve is a general requirement, one should be aware of deviations and sudden changes at certain stages, which may occur when storage areas are flooded.

The first step is to draw the curve or portions of it by visual estimation, followed by the computation of the rating curve.

*Computation of the stage discharge relation*
Assuming steady flow, mainly two types of equations are used:
1. the power equation, widely used for fitting stage discharge data

$$Q = a(h_w - h_0)^b \tag{4.44}$$

2. the square equation, more suitable for extrapolation purposes

$$Q = a(h_w - h_0)^2 + c(h_w - h_0) + d \tag{4.45}$$

where

$Q$ = discharge (m³/s)
$h_w$ = measured water level (m)
$h_0$ = water level for $Q = 0$ (m)
$a,b,c,d$ = station parameters

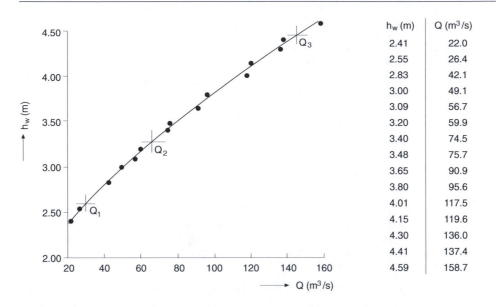

| $h_w$ (m) | Q (m³/s) |
|-----------|----------|
| 2.41 | 22.0 |
| 2.55 | 26.4 |
| 2.83 | 42.1 |
| 3.00 | 49.1 |
| 3.09 | 56.7 |
| 3.20 | 59.9 |
| 3.40 | 74.5 |
| 3.48 | 75.7 |
| 3.65 | 90.9 |
| 3.80 | 95.6 |
| 4.01 | 117.5 |
| 4.15 | 119.6 |
| 4.30 | 136.0 |
| 4.41 | 137.4 |
| 4.59 | 158.7 |

Figure 4.52. Rating curve plotted on linear scale.

In most cases the power equation is applied for the analytical determination of the rating curve or portions of it. This method consists of two steps:
1. the determination of $h_0$
2. the determination of $a$ and $b$ by log-transformation
    The method is described below and illustrated with an example.
    In Figure 4.52 a number of stage discharge measurements have been plotted and a smooth curve has been drawn by visual estimation.

To determine $h_0$ three values of discharge $Q_1$, $Q_2$ and $Q_3$ are selected from the curve so that $Q_2^2 = Q_1 Q_3$. If the corresponding values of the stages are $h_1$, $h_2$ and $h_3$, then it can be verified that, based on Equation (4.44):

$$h_0 = \frac{h_1 h_3 - h_2^2}{h_1 + h_3 - 2h_2} \tag{4.46}$$

After determination of $h_0$, the values of $Q$ plotted against $h_w - h_0$ on log-log scale will lie approximately on a straight line between shifts in control (see Figure 4.53). Thus the location of possible discontinuities can be detected and the portions of the curve to be handled separately, become clear. (In our example the curve presents one single portion).

Logarithmic transformation of Equation (4.44) yields

$$\log Q = \log a + b \log (h_w - h_0) \tag{4.47}$$

or substituting $Y = \log Q$, $a_0 = \log a$, and $X = \log (h_w - h_0)$,

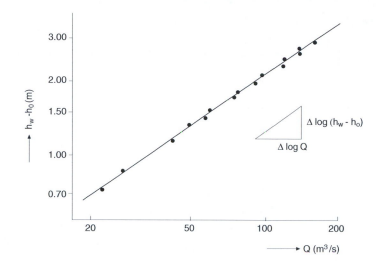

Figure 4.53. Rating curve
plotted on log-log scale.

$$Y = a_0 + bX \tag{4.48}$$

Then    $\Sigma Y = a_0 \cdot N + b \cdot \Sigma X$    $\left.\begin{array}{l} \\ \end{array}\right\}$ least squares method    (4.49)

$\Sigma XY = a_0 \cdot \Sigma X + b \cdot \Sigma X^2$    (4.50)

where $N$ is the number of data.

The parameters $a_0$ (and therefore also $a$) and $b$, are calculated using the least squares method. Substitution of the values found for $h_0$, $a$ and $b$ in Equation (4.44) then yields the mathematical model of the rating curve over the portion considered. The various portions are connected and thus the rating curve is obtained.

Note that for a wide rectangular channel $a \approx B \cdot S^{1/2}/n$ and $b \approx 5/3$ where

$B$ = width of the channel

$S$ = river slope

$n$ = Manning coefficient

In case of a strongly varying cross-section e.g. for composite cross-sections different equations for different ranges of water levels are required. The reach of each equation can easily be observed from the double logarithmic stage discharge plot by the distinct breaks in the relationship. For fitting the equations it is advisable to use an overlap between the ranges to get the intersection of the equations approximately at the required level.

*Example:*

– the file of measured stages $h_w$ and discharges $Q$ is presented in Figure 4.52

Table 4.17.Calculation of the rating curve using the least squares method.

| $Q_{meas.}$ | $h_w - 1.68$ | $y = \log Q$ | $x = \log (h_w - 1.68)$ | $xy$ | $x^2$ | $Q_{calc.}$ | $X_Q$ (%) |
|---|---|---|---|---|---|---|---|
| 22.0 | 0.73 | 1.342 | −0.1367 | −0.1834 | 0.0187 | 21.4 | −2.7 |
| 26.4 | 0.87 | 1.422 | −0.0605 | −0.0860 | 0.0037 | 27.5 | +4.2 |
| 42.1 | 1.15 | 1.624 | +0.0607 | 0.0986 | 0.0037 | 41.1 | −2.5 |
| 49.1 | 1.32 | 1.691 | +0.1206 | 0.2039 | 0.0145 | 50.0 | +1.9 |
| 56.7 | 1.41 | 1.754 | +0.1492 | 0.2617 | 0.0223 | 55.0 | −3.0 |
| 59.9 | 1.52 | 1.777 | +0.1818 | 0.3231 | 0.0331 | 61.2 | +2.2 |
| 74.5 | 1.72 | 1.872 | +0.2355 | 0.4409 | 0.0555 | 73.1 | −1.9 |
| 75.7 | 1.80 | 1.879 | +0.2553 | 0.4797 | 0.0652 | 78.0 | +3.1 |
| 90.9 | 1.97 | 1.959 | +0.2945 | 0.5769 | 0.0867 | 88.8 | −2.3 |
| 95.6 | 2.12 | 1.980 | +0.3263 | 0.6461 | 0.1065 | 98.6 | +3.2 |
| 117.5 | 2.33 | 2.070 | +0.3674 | 0.7605 | 0.1350 | 112.9 | −3.9 |
| 119.6 | 2.47 | 2.028 | +0.3927 | 0.8160 | 0.1542 | 122.8 | +2.6 |
| 136.0 | 2.62 | 2.134 | +0.4183 | 0.8927 | 0.1750 | 133.6 | −1.8 |
| 137.4 | 2.73 | 2.138 | +0.4362 | 0.9326 | 0.1902 | 141.7 | +3.1 |
| 158.7 | 2.91 | 2.201 | +0.4639 | 1.0210 | 0.2152 | 155.3 | −2.2 |
| Σ | | 27.921 | 3.5052 | 7.1843 | 1.2795 | | |

$$\left. \begin{array}{l} 27.921 = 15a_0 + 3.5052b \\ 7.1843 = 3.5052a_0 + 1.2795b \end{array} \right\} \quad \begin{array}{l} b = 1.433 \\ a_0 = 1.527 \rightarrow a_1 = 33.6 \end{array}$$

$$Q = 33.6(h_w - 1.68)^{1.433}$$

$$X_Q = 100 \frac{Q_{calc.} - Q_{meas.}}{Q_{meas.}} (\%)$$

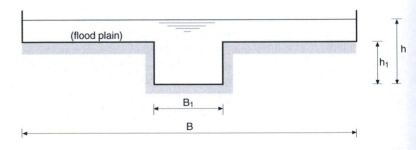

Figure 4.54. Cross-section with flood plains.

– three $Q$-values are taken from the smooth hand-drawn curve, and the corresponding stages $h_w$ are read:

$$Q_1 = 30 \text{ m}^3/\text{s} \qquad\qquad \rightarrow h_1 = 2.60 \text{ m}$$
$$Q_3 = 145 \text{ m}^3/\text{s} \qquad\qquad \rightarrow h_3 = 4.46 \text{ m}$$
$$Q_3 = \sqrt{30 * 145} = 66 \text{ m}^3/\text{s} \qquad \rightarrow h_3 = 3.28 \text{ m}$$

$$h_0 = \frac{h_1 \cdot h_3 - h_2^2}{h_1 + h_3 - 2h_2} = 1.68 \text{ m}$$

– plot the file of data ($h_w - 1.68$) and $Q$ on log-log paper.

Figure 4.53 shows one single straight line.
From Figure 4.53 it can be seen that the gradient of the rating curve is expressed by $b = \Delta\log Q/\Delta\log(h_w - h_0)$.
For prismatic channels; $1.3 < b < 1.8$.
− calculate $a$ and $b$ using the least squares method
$$\log Q = \log a + b\log (h_w - 1.68)$$

Table 4.17 shows the results of this computation.

For the measured range $20 \text{ m}^3/\text{s} < Q < 160 \text{ m}^3/\text{s}$ the rating curve under steady flow conditions is expressed as

$$Q = 33.6(h_w - 1.68)^{1.433}$$

*Flood plains*
If the flood plains carry flow over the full cross-section, the discharge consists of two parts

$$q_{river} = hB_1\, C_1\sqrt{h \cdot S} \tag{4.51}$$

and

$$Q_{flood\,plains} = (h - h_1)\, (B - B_1)\, C_2\sqrt{(h - h_1) \cdot S} \tag{4.52}$$

Assuming for the example that the flood plains have the same slope as the river bed, the total discharge becomes

$$Q = h \cdot B_1 \cdot C_1\sqrt{h \cdot S} + (h - h_1) \cdot (B - B_1) \cdot C_2\sqrt{(h - h_1) \cdot S} \tag{4.53}$$

which is illustrated in Figure 4.55. The rating curve changes significantly as soon as the flood plains are flooded especially if the ratio of the storage width $B$ to the width of the river bed $B_1$ is large.

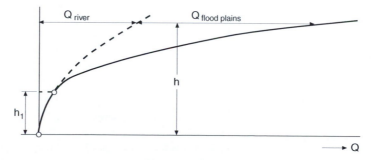

Figure 4.55. Rating curve for river with flood plains.

### 4.6.4 *Extrapolation of rating curves*

The most difficult discharges to be measured are the very low and the very high discharges. Especially the lack of measurements during peak flow is felt many times as a problem.

Various methods were developed to extrapolate the rating curve. Their applicability depends on the shape of the cross-section. If the cross-section does not change beyond the measured range, the double logarithmic straight line extrapolation can be used. Also extrapolation based on extension of area and velocity is in use.

*The log-log extension*
The rating curve, established from existing measurements, is determined using Equation (4.52):

$$Q = a(h_w - h_0)^b \quad \text{(see Section 4.6.3)}$$

After calculation of $h_0$ with Equation (4.46) the $Q - (h_w - h_0)$ relation is plotted on a log-log scale. Possible discontinuities are made visible, and the portions to be handled separately become clear.

The portion $Q = a(h_w - h_0)^b$ representing the highest present measurements is extrapolated.

*Extension of area and velocity (see Figure 4.56)*
Based on the existing measurements $h_w$ and $Q = \bar{v} \cdot A$, the mean velocity as well as the crosssectional area $A$ are plotted against stage $h_w$.

The curve $h_w - A$ is drawn and extrapolated by hand, taking into account possible irregularities as flood plains.

The curve $h_w - \bar{v}$ is drawn and extrapolated using the Manning equation $v = \dfrac{1}{n} \cdot R^{2/3} \cdot S^{1/2}$ (taking for $n$ and $S$ constant values based on the existing measurements).

The discharge for an extrapolated stage $h_e$ becomes $Q_e = \bar{v}_e \cdot A_e$

*Extension of conveyance and slope*
When cross-sectional changes (river with floodplains) and different velocities are to be expected in the cross-section, the Manning or Chézy equation is to be applied to each part.

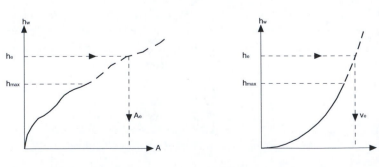

Figure 4.56. Extension of area and velocity with water depth $h$.

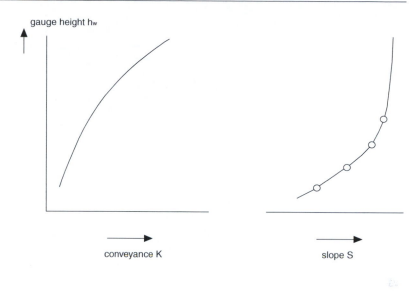

Figure 4.57. High flow
extrapolation using the
conveyance slope method.

In the conveyance slope method, see Figure 4.57, the conveyance and
the energy slope are extrapolated separately.

The Manning equation reads:

$$Q = \frac{1}{n} \cdot R^{2/3} \cdot S^{1/2} \cdot A \qquad (4.54)$$

$$Q = K \cdot S^{1/2}$$
where the conveyance is $K = A \cdot R^{2/3}/n$ $\qquad (4.55)$

Both $K$ and $S$ are functions of the measured gauge height $h_w$.

The conveyance $K$ can be extrapolated based on the measured cross-
sectional values $A$ and $R$.

The energy gradient $S$ is extrapolated by extending the measured
range and keeping in mind that $S$ tends to become constant at the higher
stages: the bed slope of the river.

### 4.6.5 Non-unique rating curves

A non-unique $Q$-$h$ relationship may be caused by one of the following
phenomena:
- rising or falling stages (flood wave)
- changes in cross-sectional area (morphological changes)
- changes in vegetation
- backwater effects

*Rising or falling stages*
From Chézy's equation $Q = \int_o^B C\sqrt{S} \cdot h^{3/2} \cdot dB$ it follows that a change in
the water level gradient will also result in a change of the discharge.

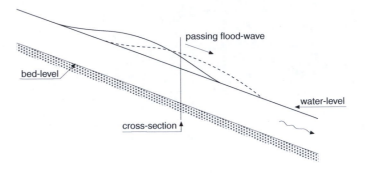

Figure 4.58. Flood wave travelling down a river (schematically).

Generally, the water level gradient becomes steeper when the discharge increases, because a greater (permanent) discharge implies a greater water depth. The reverse occurs with decreasing discharges. Hence the passing of a flood wave, resulting in a water level rise and fall, leads to higher discharges in the rising stage and lower discharges in the falling stage.

When plotting the relationship between the water level gradient and the stage of the river, the so-called stage slope curve, it will be clear that such a curve will show a loop. When the water level gradient is calculated from the difference in head divided by the distance between two gauging stations, the loop in the stage slope curve becomes more pronounced because the travelling time of the flood wave is also included. The loop becomes still more marked when between the two gauging stations storage areas are filled and emptied by the passing wave. In view of this, it will be clear that along those rivers and channels the water level gradient has to be measured near the discharge measuring cross-section.

Figure 4.59 shows the stage slope curves for an upper river and a lower river.

The loop in the stage slope curve also implies a loop in the relationship between the stage of the river and the discharge: the so-called stage discharge curve or rating curve. In view of the fact that the water level gradient ($\partial h/\partial x$) already decreases when the stage of the river still increases (Figure 4.58), the maximum discharge will occur before the maximum stage of the river is reached (Figure 4.60).

Figure 4.59. Schematic stage slope curves for main rivers.

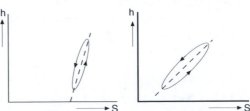

a. Upper main river          b. Lower main river

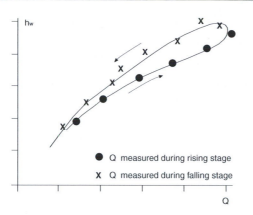

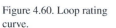

Figure 4.60. Loop rating
curve.

The rising and falling stages are originally caused by the flood wave
(a) and further they may be intensified by (b) changes in the bottom
ripples (sand dunes in alluvial streams) and (c) storage effects, downstream of the measuring section.

a) Mostly the front of a flood wave is steeper than the back of the wave.
   Due to steeper energy lines the water level in a section will – for a
   certain discharge – be lower during the rising stage than during the
   falling stage.
b) During a steep rise of water levels bottom ripples tend to grow, but
   often the forming of bigger ripples lacks somewhat behind in time.
   This means that during the rising stage this roughness effect leads to
   too-smooth bottoms (compared with an equilibrium situation), so to
   relatively too-low water levels.
c) Storage effects. If downstream of the measuring section storage (overflow of banks) occurs, for a certain $Q$ the backwater effect will cause
   relatively lower water levels during rising stages.

The effect of:
– relatively steep front of flood wave
– time lag in adaption of ripple roughness
– downstream storage
work in the same direction, leading to a $Q$-$h_w$ curve as shown in Figure 4.60.

The unsteady flow $Q_m$ in the loopings for the rising and the falling stages
can be estimated as follows:

$$Q_m = Q \cdot \sqrt{1 \pm \frac{\dfrac{dh}{dt}}{S_c \cdot v_w}} \qquad (4.56)$$

where
$Q_m$ = unsteady flow (m$^3$/s)
$Q$   = steady flow according to the rating curve (m$^3$/s)
$\dfrac{dh}{dt}$ = rate of change of the water level (m/s)

$S_c$    = bottom slope ≈ energy slope for steady flow (−)
$v_w$    = celerity of the flood wave (m/s) $v_w$ = 1.5 v
v      = mean flow velocity in the cross-section (m/s)

*Example:*
$S_c = 1.5 * 10^{-4}$
$v = 1.10$ m/s → $v_w = 1.65$ m/s
$\dfrac{dh}{dt} = 0.30$ m in 1 hour → $\dfrac{dh}{dt} = 8,33.10^{-5}$ m/s
then $\dfrac{Q_m}{Q}$ ranges from 0.81 to 1.16

To apply Equation (4.56) either $S_c$ and $v_w$ should be known or sufficient data must be available to estimate the adjustment factor $1/S_c v_w$ as a function of the gauge height.

*Changes in cross-sectional area*
Sedimentation or erosion over a great length of a river – morphological changes – results in a temporary accretion or degradation of the bottom level, leading to a temporary rise or fall of the water level.
    The same level fall happens after cleaning irrigation canals and drains.

*Changes in vegetation*
Vegetation – seasonal changes – generally reduces the cross-sectional area and results in a higher roughness of bottom and banks. The joint effect is often expressed in a varying Manning factor *n*.
    Ven Te Chow gives for natural streams, $B < 30$ m the values indicated in Table 4.18.

*Backwater effects*
–  tidal motion on the lower course of rivers
–  influence of flow distribution in confluences

*Conclusions:*
–  A once-established unique rating curve $Q\text{-}h_w$ can only be used accurately, as long as none of the non-unique phenomena – floodwave, morphological changes, changes in vegetation or back water effects

Table 4.18. Variation of roughness coefficient *n*.

| | | | |
|---|---|---|---|
| 1. Clean, straight, full stage, no rifts or deep pools | 0.025 | 0.030 | 0.033 |
| 2. Same as above, but more stones and weeds | 0.030 | 0.035 | 0.040 |
| 3. Clean, winding, some pools and shoals | 0.033 | 0.040 | 0.045 |
| 4. Same as above, but some weeds and stones | 0.035 | 0.045 | 0.050 |
| 5. Same as above, lower stages, more ineffective slopes and sections | 0.040 | 0.048 | 0.055 |
| 6. Same as 4, but more stones | 0.045 | 0.050 | 0.060 |
| 7. Sluggish reaches weedy, deep pools | 0.050 | 0.070 | 0.080 |
| | ← | variation | → |

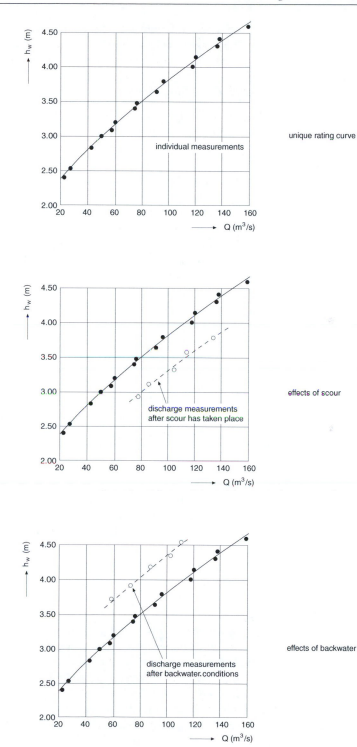

Figure 4.61. The effects of scour and backwater on the unique rating curve.

are present. Figure 4.61 shows the effects of scour and backwater.
–  The stage discharge method is a reliable method, provided check measurements are carried out at regular intervals (at least ten times per year).

### 4.6.6  *Processing of stage discharge data*

Since about 1980 various software packages have been developed to store, manage and process stage discharge data.

Processing of data is done in the following steps:
–  Elaboration of discharge measurements. Example: in the velocity area method the basic data, depths and flow velocities are elaborated, using the mean section method or the mid section method
–  The stage discharge data base
   The following data and parameters of each discharge measurement should be stored in the stage discharge data base:
   –  number of the measurement
   –  date and time of the measurement
   –  water level
   –  discharge
   –  rate of change of the water level
   –  width of the river at the water line
   –  cross-sectional area, and
   –  wetted perimeter
   Each measurement includes a record of the data base with the above parameters for a particular station. The data base is used for further analysis of the measurements.
–  Computation of the rating curve
   steady flow: computation of the unique rating curve, in most cases using the power type equation $Q = a \cdot (h_w - h_0)^b$ (Section 4.6.3)
   unsteady flow: see Section 4.6.5
–  Validation of rating curves
   To investigate the validity of the rating curve when new measurements become available, tests for absence from bias and goodness-of-fit can be applied, such as a run test on the sequence of positive and negative deviations from the rating curve.

Daily flow values should be based on 15 minutes values. Where the stage discharge method is applied, the 15 minutes water level values are converted in 15 minute discharge values, from which hourly, daily, monthly and annual values are calculated.

The 15 minute values are stored in the data base for statistical evaluation.

Figure 4.62 shows the daily discharges for a gauging station in the River Roer (the Netherlands).

### 4.6.7  *Statistical evaluation of discharge data*

Many other curves can be derived from the rating curve. Some examples are given.

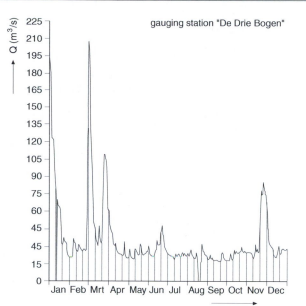

Figure 4.62. Discharge hydrograph River Roer for 1987.

*Relation curves*

When the stage discharge curve for consecutive gauging stations along the same river have been drawn, relation curves between the water levels at these stations can be established. However, the assumption must be made that the discharge between two stations is not modified by, e.g. tributaries or storage areas.

A schematic example is given in Figure 4.63.

However, stage discharge curves are not unique relations due to the scatter of the measured discharges around the average. To prevent this scatter, permanency of the discharge should be assumed. The assumption made (no modification of the discharge between two stations occurs) will, generally, never hold in practice. Nevertheless, water level relations can sometimes still be established. The water levels actually observed should

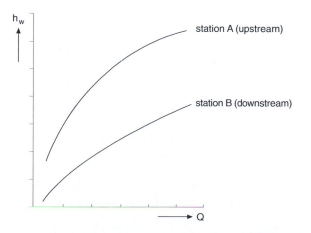

Figure 4.63. Discharge relation curve (after: Nedeco, 1973).

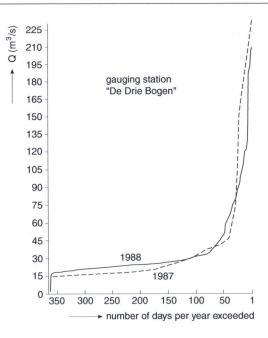

Figure 4.64. Flow duration curve River Roer for 1987 and 1988.

then be smoothed due to the disturbances by the non-permanency of the discharge.

*Duration curves*
Arrangement of the observed data in order of ascending or descending magnitudes allows determinations of the percentage of the time (or the number of days per year) that a certain value of discharge is equalled or exceeded. Then the magnitudes plotted against the corresponding percentages (or days) result in an empirical *duration curve*. The derivative of such a curve at the magnitude under observation gives the probability density.
   The figures 4.64 and 4.65 are examples of duration curves.

Figure 4.64 shows the flow duration curves for the years 1987 and 1988, in the gauging station 'De Drie Bogen', River Roer. The curve gives the number of days that a certain discharge has been lower than indicated by the curve.

The duration curve deduced from observations over a period of many years – a whole number of years in view of the seasonal variations – enables a reasonable estimate of the probability that a certain discharge (or water level) is equalled or exceeded; in other words the part of the year that this happens on average.
   Figure 4.65 shows the duration curves and the long term average for the River Rhine in the Netherlands.

The records used for the drawing of these curves can be considered as samples taken from a long history of flow variations. The greater the

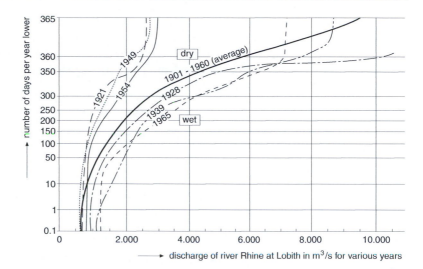

Figure 4.65. Duration curves
(years) for the River Rhine
(after: Jansen, 1979).

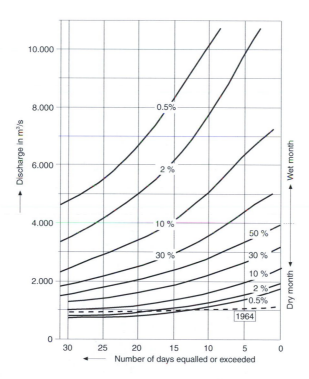

Figure 4.66. x% wet and dry
month (after: Jansen, 1979).

sample (i.e. the longer the period of observations), the more accurately
the character of the relationship will be approximated. The empirical
duration curve of one year, therefore, usually deviates markedly from the
average (see Figure 4.65).

*Wet and dry periods*
For engineering purposes, it is of interest to know the probability of
occurrence of the deviations. For practical use, curves for x% wet and dry

years, or months, can be deduced from data covering a sufficiently long observation period (i.e. 50 years). Such a curve is shown in Figure 4.66 for wet and dry months. Obviously the 50% curve represents the median. From such sets of curves the regime of a certain year or month can be evaluated.

In Figure 4.66 the duration curve of January 1964 is indicated, enabling evaluation of the degree of dryness of that particular month (dotted line ---- 1964 in the figure).

For a full account of the theory and application of statistical methods the reader is referred to Ven Te Chow (1964), Sturm (2001), WMO (1970) and other handbooks.

## 4.7  SLOPE STAGE DISCHARGE METHOD

In some cases it is impossible to select a discharge gauging station in a reach which is free from backwater effects:
- regulated water courses
- confluences
- variable water of a downstream reservoir
- downstream constrictions with a variable capacity (weed, ice, etc.)
- rivers with return of overbank flow (flood plains)

The presence of variable backwater does not allow the use of a simple unique rating curve. Variable backwater causes a variable energy slope at a given stage. Now the discharge is a function of both stage and slope.

Many of these sites can be operated using the slope stage discharge method, also referred to as the stage fall discharge method (ISO).

The stage is measured continuously using a base gauge. An auxiliary gauge some distance downstream from the base gauge, and set to the same datum, is also recording continuously. Assuming a precise time synchronisation between both gauges, the water surface slope can be calculated, from which records of energy slope are obtained.

The length of the reach between both water level gauges should be such that a fall $F$ of at least 0.15 m is measured.

Basically, the discharge is derived using the Manning equation.

$$Q = \frac{1}{n} \cdot R^{2/3} \cdot S^{1/2} \cdot A$$

where $R$ and $A$ are average section parameters.

In practice, discharges are calculated using one of the following methods:
- estimation of a Manning coefficient $n$, and calculation of discharges with the measured values of $R$, $S$ and $A$.
- the constant fall rating (recommended where backwater is always present)

– the variable fall rating (intermittent backwater conditions)

The latter two methods are described in an ISO technical report, TR 9123. A specific application of the constant fall rating – the unit fall rating – is described below.

The unit fall method is used with the assumption that the relation between the discharge ratio $Q/Q_n$ and the fall ratio $F/F_n$ is a square root relation, given by the following equation.

$$Q/Q_n = \sqrt{F/F_n} \tag{4.57}$$

where
$Q$ = measured discharge current-meter method, (m³/s)
$F$ = measured fall (m)
$Q_n$ = discharge from the rating curve corresponding to the base gauge water level (m³/s)
$F_n$ = the constant fall (m)

In the unit fall method $F_n = 1.00$ m. Equation (4.57) becomes:

$$Q_n = Q/\sqrt{F} \tag{4.58}$$

The rating curve is developed by plotting the measured $Q/\sqrt{F}$ values against the corresponding measured water level $h$ from the base gauge. The rating curve is then fitted to the plotted points.

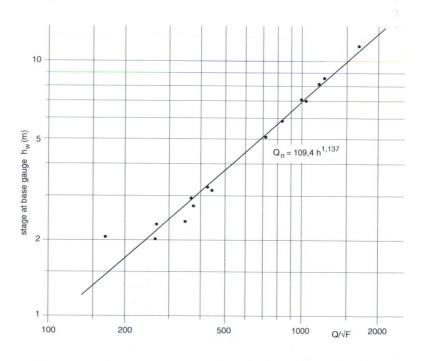

Figure 4.67. Unit fall rating curve for $F_n = 1.00$ m (after: ISO).

Table 4.19. Unit fall rating (after: ISO).

| Meas. no. | Measurements | | | | Calculation of difference $X_Q$ | | |
|---|---|---|---|---|---|---|---|
| | Stage at base gauge $h$ (m) | Stage at auxiliary gauge $h_a$ (m) | Fall $F = h - h_a$ (m) | Discharge $Q$ m³/s | $Q_n = Q / \sqrt{F}$ | $Q_n = 109.4 h^{1.137}$ m³/s | $X_Q$ % |
| 1 | 2.012 | 1.951 | 0.061 | 66.0 | 267 | 242 | −9.4 |
| 2 | 2.036 | 1.978 | 0.058 | 39.9 | 166 | 246 | 48.2 |
| 3 | 2.286 | 1.996 | 0.290 | 145 | 269 | 280 | 4.1 |
| 4 | 2.359 | 2.155 | 0.204 | 156 | 345 | 290 | −15.9 |
| 5 | 2.755 | 2.054 | 0.701 | 317 | 379 | 346 | −8.7 |
| 6 | 2.963 | 2.347 | 0.616 | 289 | 368 | 376 | 2.2 |
| 7 | 3.139 | 2.331 | 0.808 | 399 | 444 | 402 | −9.5 |
| 8 | 3.206 | 2.279 | 0.927 | 411 | 427 | 411 | −3.5 |
| 9 | 5.026 | 3.429 | 1.597 | 889 | 703 | 686 | −2.4 |
| 10 | 5.907 | 3.990 | 1.917 | 1160 | 838 | 826 | −1.4 |
| 11 | 7.013 | 4.788 | 2.225 | 1490 | 999 | 1002 | 0.3 |
| 12 | 7.105 | 4.923 | 2.182 | 1520 | 1029 | 1017 | −1.2 |
| 13 | 8.108 | 6.188 | 1.920 | 1640 | 1184 | 1182 | −0.2 |
| 14 | 8.638 | 5.986 | 2.652 | 1990 | 1222 | 1270 | 3.9 |
| 15 | 11.558 | 8.678 | 2.880 | 2830 | 1668 | 1768 | 6.0 |

Table 4.19 shows an example of a unit fall rating for a site with high backwater from a power dam where backwater exists at all stages and at all times: measurement of stages and discharges, resulting in the rating curve $Q_n = Q / \sqrt{F} = 109.4 \, h^{1.137}$ (see also Figure 4.67).

From this curve, discharges are computed for any given base gauge water level $h$ by taking the $Q_n$ from the curve and multiplying this value by the square root of the respective measured fall $F$.

$$Q = Q_n \cdot \sqrt{F} \qquad (4.59)$$

*Example:*

stage $h = 2.50$ m $\rightarrow Q_n = 109.4 * 2.50^{1.137} = 310.1$ m³/s

fall $F = 0.44$ m $\rightarrow Q = 310.1 * \sqrt{0.44} = 206$ m³/s

More information about the constant fall and the variable fall ratings is given in ISO/TR 9123.

## 4.8 ACOUSTIC METHOD

The principle of the acoustic (ultrasonic) method is the measurement of the flow velocity at a certain depth in the channel by transmitting acoustic or sound pulses through the water from transducers located on the banks on both sides of the river.

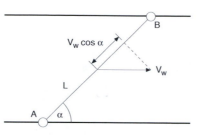

Figure 4.68. Principle of the acoustic method.

These transducers are installed on either side of the water course below the water surface at a certain depth and are positioned in such a way as to create a diagonal acoustic measuring line.
Acoustic pulses are being transmitted from A to B and vice versa through the river by acoustic transducers (see Figure 4.68).

With zero flow the travel time from A to B ($t_1$) is equal to the travel time from B to A ($t_2$).

$$t_1 = t_2 = \frac{L}{C} \tag{4.60}$$

where
$L$ = the length of the acoustic path (m)
$C$ = the velocity of sound in water, 1450–1480 m/s

If there is a flow as in Figure 4.68, then the acoustic pulse from A to B is being accelerated and the acoustic pulse from B to A is being delayed, according to:

$$t_1 = \frac{L}{C + v_w \cos \alpha} \quad \text{travel time in downstream direction} \tag{4.61}$$

and

$$t_2 = \frac{L}{C - v_w \cos \alpha} \quad \text{travel time in upstream direction} \tag{4.62}$$

where
$v_w$ = the mean velocity vector of the flow at the depth of the transducers
$\alpha$  = the angle between this vector and the acoustic path
From the difference in the travel time, the flow velocity $v_w$ can be deduced.

$$v_w = \frac{L}{2 \cos \alpha} \left( \frac{1}{t_1} - \frac{1}{t_2} \right) \tag{4.63}$$

In fact it is possible to measure the flow velocity $v_w$ using only one single diagonal acoustic measuring line, formed by two transducers installed on

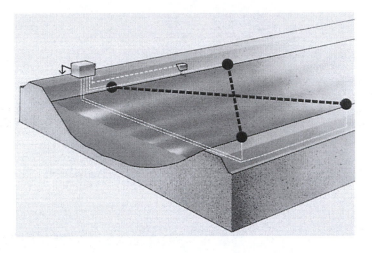

Figure 4.69. Acoustic flow measuring system with a measuring cross (after: Instromet).

either side of the watercourse. In practice, however, the influence of the wind, nearby bends and other factors cause cross flow components that should not be neglected. By adding a second measuring line across the first line, a measuring cross is formed: this completely eliminates the influence of the cross flow and produces a correct flow velocity $v_w$.

In some cases – deep water and a big difference between the maximum and the minimum water level – two measuring crosses are installed at different depths.

In most cases only one measuring cross will suffice.

For a correct calculation of the real average flow velocity v in the cross-section, the measured velocity $v_w$ needs to be corrected with a factor $K$, which depends on:
– the depth of the measuring cross
– the bottom roughness
– the shape of the cross-section

The factor $K$ is based upon the assumption of a parabolic velocity distribution in the vertical

Normally $K$ varies as follows $0.9 < K < 1.1$. Sometimes it is recommended to derive $K$ from single measurements in the cross-section.

$$\bar{v} = K \cdot v_w \tag{4.64}$$

Since the cross-sectional area $A$ is related to the measured water level the discharge is calculated as follows:

$$Q = K \cdot v_w \cdot A \tag{4.65}$$

Using the acoustic method, both positive and negative discharges can be measured.

The application of an acoustic flow meter may be limited in the following ways:
– disturbance of the acoustic path by temperature gradients, salinity gradients, fluctuations in sediment concentration and by air bubbles. As in most systems, both acoustic pulses are transmitted at the same time, and while travelling through the same water, the velocity of sound is identical for both pulses. Therefore, the measured flow velocity is not affected by changes in the velocity of sound.
Sound is reflected from the water surface and, to a lesser extent from the channel bed. The bed is usually a net absorber of sound. Errors in signal timing will be prevented if the depth of water above the acoustic path exceeds

$$D_{min} = 27\sqrt{L/f} \tag{4.66}$$

where:

$D_{min}$  is the minimum depth of water above the acoustic path and also the minimum clearance between the bed and the path (in metres)
$L$   is the path length (in metres)
$f$   is the transducer frequency (in hertz)

*Example:*
  river width $B = 80$ m
  path length $L = 80\ \sqrt{2} = 113$ m
  transducer $f\ = 200$ kHz
  $D_{min}$     $= 0.54$ m
  Waterdepth $d > 1.08$ m

The unreliability in the determination of flow is given by:

$$X_Q = \sqrt{X_A^2 + X_L^2 + X_\alpha^2 + X_K^2} \tag{4.67}$$

where
$X_A$ = error in determining the cross-sectional area (%)
$X_L$ = error in path length (%)
$X_\alpha$ = error in angle $\alpha$ (%)
$X_K$ = error in the correction factor $K$ (%)
  The error $X_K$ is increasing for low flow velocities, v < 0.05 or 0.10 m/s where irregular velocity profiles can be developed, which would not resemble the classical parabolic velocity profiles.
  A more detailed description of the uncertainty in the discharge estimation is given in ISO 6416 (2004).

## 4.9  THE ELECTROMAGNETIC METHOD

The electromagnetic method using a full-channel-width coil is a modern method for measuring discharges in open channels. The method was developed and brought in operation for the first time in United Kingdom, and is very suitable in rivers where weed growth is an ever-lasting problem.

*Principle*
The principle of operation is as follows. The motion of water flowing in an open channel cuts a vertical magnetic field which is generated using a large coil buried beneath the riverbed, through which an electric current is driven. An electromotive force (e.m.f.) is induced in the water and measured by signal probes (electrodes) at each side of the channel. This very small voltage is directly proportional to the average velocity of flow in the cross-section.

Figure 4.70 gives a diagram of the system.

Faraday's law (1832) of electromagnetic induction relates the length of a conductor moving in a magnetic field to the e.m.f. using the following basic relationship

$$E = v \cdot b \cdot B \tag{4.68}$$

where
$E$ = electromotive force in volts (V)
$v$ = average velocity of the conductor (water) in m/s
$b$ = length of the conductor, equal to the channel width in metres
$B$ = magnetic field intensity in teslas (t)

In practice however, the theoretical output potential from the signal probes will be reduced depending on the bed and water conductivity. Hence Equation 4.68 is transformed to:

$$Q = K \cdot \left( \frac{E \cdot R_w}{I \cdot R_b} - C \right)^n \tag{4.69}$$

where
$Q$  = discharge (m³/s)
$K$  = dimensional constant
$R_w$ = water resistivity (Ω/m)
$R_b$ = bed resistance (Ω)
$I$   = coil current (A)
$C$  = station constant (V/mA)
$n$  = exponent

The values of $K$, $C$ and $n$ are empirical values, following from the calibration of the system.

The shortening effect of the bed may be reduced by lining the channel with an electrically insulating impervious membrane to reduce the current-leakage to an acceptable level.

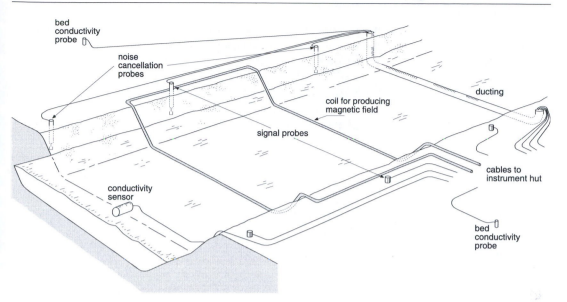

bed
conductivity
probe

noise
cancellation
probes

ducting

coil for producing
magnetic field

signal probes

cables to
instrument hut

conductivity
sensor

bed
conductivity
probe

Figure 4.70. Diagram of an
electromagnetic gauging
station (after: Herschy, 1978).

*Applications*

The electromagnetic method is particularly suited in the following cases:
– flow measurement of untreated domestic effluent, treated effluents,
  potable water in treatment works and cooling water in power stations. In all these cases the channel will be rectangular and made of
  concrete
– flow measurement in rivers with weed growth, high sediment concentrations, unstable bed conditions, backwater effects or reverse flow.

*Selection of site*
– areas of high electrical interference (power cables, electric railways)
  should be avoided
– for non-insulated channels the width/depth ratio shall not exceed
  10. Insulated channels allow a width/depth ratio of 200, according to
  the standard ISO 9213.
– spatial variation in water conductivity should be avoided (many
  estuaries)

*Design and construction* (see Figure 4.70 reproduced with permis-
sion). The electromagnetic gauging station is composed of the following
elements:
– the coil buried under the channel at a depth of 0.5 m, or bridging
  across the channel
– an insulating membrane, normally necessary
– two signal probes (electrodes) measuring the output potential
– two probes measuring the bed conductivity
– four noise cancellation probes situated outside the magnetic field,
  measuring the surrounding electrical noise, to be deducted from the
  signal produced by the voltage probes

– the maximum width depends on the costs of instrumentation and installation. In most cases the width of the river shall not exceed 50 to 100 metres.

*Calibration*

The electromagnetic gauging station requires an on-site empirical calibration, using one of the incidental methods (mostly the current-meter method).

The rating curve is based on the coil configuration and the measured range of stage. The curve can be in the form of Equation 4.69:

$$Q = f\left(\frac{E \cdot R_w}{I \cdot R_b}\right)$$

*Uncertainty*

– in insulated channels of rectangular cross-section, the total uncertainty at the 95% confidence level will be about 5%
– in non-insulated channels, the total uncertainty will be about 10%
    Using the electromagnetic method, minimum velocities of 0.002 m/s can be detected.

## 4.10  PUMPING STATIONS

### 4.10.1 *Introduction*

Determination of the water volumes passing a pumping station can be done by registration of the operation-hours, by measuring both water levels: the supply level and the outlet level, and using the so-called pump curve.

    Roughly two main families of pumping stations are distinguished:
– the Archimedean-type screw pump, where water is lifted through an open channel (Section 4.10.2)
– centrifugal (and other) pumps, where water is transported through closed pipes (Section 4.10.3)

### 4.10.2 *The Archimedean-type screw pump*

*Origin*

For the origin of the Archimedean-type screw pump, which is a conveyor rather than a real pump, one has to go back to Archimedes. He is considered as the inventor of the system of lifting water using a rotating spiral. Old pictures, however, do not show the screw pump in the modern shape in which it is used nowadays, but a closed spiral (Figure 4.71). In this form it is still used in some eastern countries.

*Construction*

The Archimedean-type screw pump has one or more helically wounded blades, attached to a shaft (pipe) thus forming one or more open spirals in

Figure 4.71. Original screw pump as a closed spiral.

which water can be held if the screw is placed in an accurately fitting trough. Mounted under a certain gradient, it will lift its content when rotated.

*Comparison with real pumps*
Comparison with for instance centrifugal pumps shows the following advantages. The screw pump works at a low speed, hence the wear is negligible. The screw is open, so the process is surveyable. The spirals are large compared to the passages of pumps, which is of great importance in the case of lifting crude sewage. Of course there are some disadvantages too. Low speed means small velocities and consequently big dimensions. The head is limited on account of the length of the screw. Pressure piping can not be applied.

*Head and capacity*
A head of 10 meters at one stage is within the reach of the modern screw pump. For greater heads there are solutions as intermediate bearing and two-stage installations. Capacities of 5 m³/sec can be reached.

The capacity is proportional to the quantity of water between two blades, multiplied by the number of revolutions $n$ or, if the screw has more spirals, once more multiplied by the number of spirals.

Figure 4.72 gives a principle sketch of the modern screw pump.

For the modern Archimedean screw pump the following symbols and notions are defined:

$d$ = diameter of the shaft (m)
$D$ = diameter of the screw (m)
$\beta$ = gradient, in most cases 30°, 35° or 38°
$a$ = number of spirals, in most cases three (–)
$s$ = pitch of an individual spiral (m)
$L$ = total length of spirals along the shaft (m)

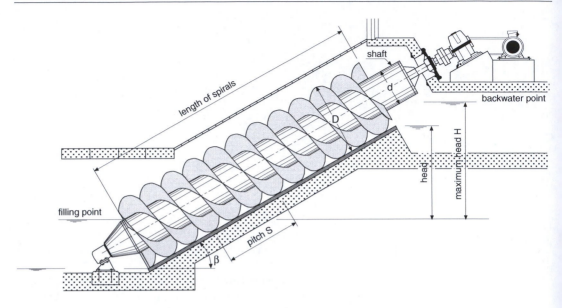

Figure 4.72. Layout of
a modern Archimedean
screw pump (after: Jansen
Venneboer).

trough:    half-cylindrical channel in which the screw rotates.
          The trough can be manufactured of steel or in concrete.

The intersection of the water levels on the screw with the blades gives the
points *A* and *B* to which the trough must enclose the screw (Figure 4.73).
These points must not be higher than diametrical to each other in order
to make the mounting of the screw possible. On the side of the downward movement a large angle α prevents solids from entering between
the screw and the trough. At the other side, a small angle *β* prevents
leakage in case the water levels in the screw are a slightly higher than the
theoretical ones.

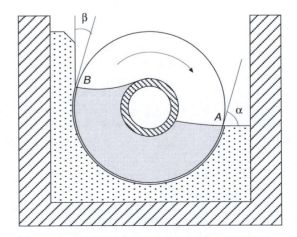

Figure 4.73. Cross-section
over the trough.

filling point:    supply level leading to maximum efficiency
backwater point:  outlet level leading to maximum efficiency (maximum
                 head). This point is situated 0.15 to 0.33 $D$ above the top of
                 the trough (depending on the parameters $\beta$, $d/D$, $S/D$ and $a$).
The maximum capacity occurs for the following conditions:
–   supply level ≥ filling point
–   outlet level ≤ backwater point
The maximum efficiency occurs for:
–   supply level = filling point
–   outlet level = backwater point
    If the outlet level is too high, then the water will flow back over the
screw.
    Figure 4.74 shows capacity and efficiency versus supply level. The
discharge through the Archimedean screw pump – under maximum
capacity conditions – is given by Equation 4.70:

$$Q = f \cdot q \cdot n \cdot D^3 \tag{4.70}$$

where
$Q$ = discharge (m³/minute)
$q$  = screw factor (–), $q = f(\beta, d/D, S/D)$
        values of $q$ are given in literature or by the manufacturer
        Example: for $\beta = 30°$, $d/D = 0.5$ and $S/D = 1.0$ follows $q = 0.250$
$f$ = efficiency factor (–), normally $1.10 < f < 1.20$
        $f = 1.10$   old installations
        $f = 1.15$   average value
        $f = 1.20$   new installations, made with care and skill
$n$  = speed (revolutions per minute), normally $20 < n < 100$
        For design purposes $n = 50/D^{2/3}$ ($D$ in metres)    (4.71)
$D$ = screw diameter (m)

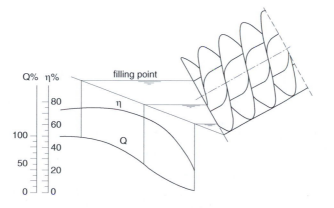

Figure 4.74. Capacity and
efficiency depending on the
supply level (after: Spaans).

*Uncertainties*

The total random error in estimating the discharge is expected to be $X_Q =$ 3 to 6%.

Systematic errors will occur if the supply level and the outlet level do not fulfil the above mentioned conditions for the maximum capacity.

### 4.10.3  *Centrifugal pumps*

Centrifugal pumps (radial flow) are the most common type of pumps used in irrigation schemes.

Figure 4.75 shows a centrifugal pump.

Centrifugal pumps form part of closed pipe systems. Water supply takes place via a suction pipe, the outlet is a delivery pipe.

The determination of discharges through centrifugal pumps is more complicated.

For centrifugal pumps the following symbols and notions are defined:

$H_{man}$ = manometric head, difference in pressure at a short distance before and behind the pump, measured with manometers

$H_{pump}$ = difference in energy head at a short distance before and behind the pump

$$H_{pump} = H_{man} + z + \frac{v_1^2}{2g} - \frac{v_2^2}{2g}$$

where

z  = difference in level between both manometers

$v_1$ = flow velocity in the suction pipe

$v_2$ = flow velocity in the delivery pipe

$H_{stat}$ = difference between the open water levels in the reservoir or channel at both sides of the pumping station

Figure 4.75. A centrifugal pump (after: Kay, 1992).

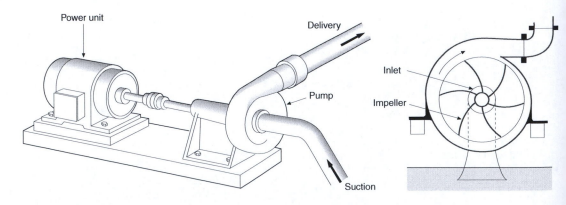

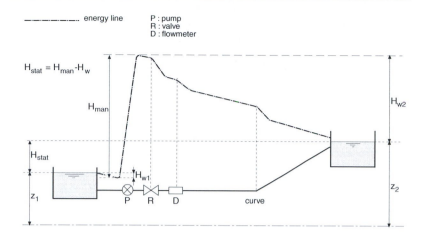

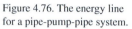

Figure 4.76. The energy line
for a pipe-pump-pipe system.

$H_w$    =   total energy losses, $H_{w1}$ before and $H_{w2}$ behind pump $H_w = H_{w1} + H_{w2}$. This total energy loss includes losses by friction and losses caused by inflow, curves, valves, trashracks, outflow, etc.

Figure 4.76 gives the energy line for a pipe-pump-pipe system.

In the factory $H_{pump}$ is measured as a function of the discharge $Q$.
If we assume:
$z = 0$ (both manometers at the same level)
$v_1 = v_2$ (diameters suction pipe and delivery pipe are the same)
then    $H_{pump}$   = $H_{man}$   and
          $H_{man}$   = $H_{stat} + H_w$

Figure 4.77. Factory test
according to ISO 2548
(after: ISO).

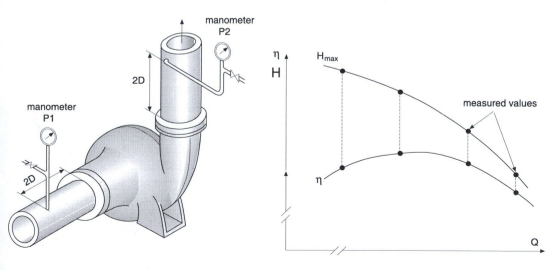

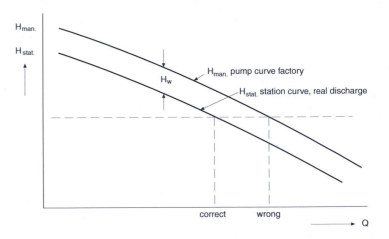

Figure 4.78. Bad practice of using the pump curve.

Now two types of discharge curves should be distinguished:

1. Pumpcurve:    $Q - H_{man}$ This relation has been set up in the factory Figure 4.77 shows the factory calibration: capacity and efficiency $\eta$

2. Stationcurve:    $Q - H_{stat}$ This relation has been set up using the pump curve, and by computation of the $H_w$ losses, or by calibration in the field (current-meter method)

Determination of the discharges $Q$ passing the pumping station can be done in one of the following ways:

a) using the pump curve $Q - H_{man}$ and measuring $H_{man}$ at the same locations as in the factory test.

b) using the pump curve $Q - H_{man}$, but measuring $H_{stat}$ (neglecting $H_w$). This method is common practice in many cases, leading to a systematic overestimation of the real discharges, depending on the ratio $H_w/H_{man}$. See Figure 4.78.

c) using the station curve $Q - H_{stat}$ after calibration of the pumping station, and measuring the open water levels at both sides of the station

d) measuring the discharges in the delivery pipe using an electro-magnetic flow meter, ultra sonic sensors or doppler flow meters. This method is attractive if several pumps are discharging in one or a few delivery pipes.

e) measuring the discharges in one of the open channels, suction side or delivery side, using the acoustic method or using a flow measurement structure (a weir or flume).

## 4.11  FLOW MEASUREMENT STRUCTURES

The use of flow measurement structures – weirs, flumes and gates – is one of the various methods for the continuous measurement of discharges in open channels.

Chapter 6 is dealing with flow measurement structures, their functions, the classification of types, the fields of application and the selection of the most appropriate type.

## 4.12 SELECTION OF DISCHARGE MEASUREMENT METHODS

The following factors may affect the selection of the discharge measurement method:
– required accuracy
– availability of equipment (vehicles and instruments)
– availability and skill of personnel, local experience
– accessibility of the site and the stream
– costs
– width and depth of the stream
– range of flow velocity
– frequency of measurements
    The limiting conditions of the methods described in this chapter are given in Table 4.20.
    The first eight methods are single measurement methods, the remaining seven are continuous methods.

Table 4.20. Limiting conditions of discharge measuring methods.

| Method | | Criteria | | | | | | | Uncertainty | |
|---|---|---|---|---|---|---|---|---|---|---|
| Description | Relevant Intern. Standard | Width | Depth | Velocity | Sediment load | Approach conditions | Time factor | Minimum percentage (%) | Comment |
| Velocity area, by wading | ISO 748 | S,M,L | S | S,M | | b,c,d | J,K | 3 to 6 | A,B |
| Velocity area, from bridge | ISO 748 | M,L | M,L | M,L | | b,c,d | K | 3 to 6 | A,B,C,D |
| Velocity area, cableway | ISO 748 | M,L | M,L | M,L | | b,c,d | K | 3 to 6 | A,B,C |
| Velocity area, static boat | ISO 748 | M,L | M,L | M,L | | b,c,d | K | 3 to 6 | A,B,C,E |
| Velocity area, moving boat | ISO 4369 | L | M,L | M,L | | b,c,d | K | 5 to 10 | A,B,E |
| Velocity area, floats | ISO 748 | M,L | M,L | S,M,L | | b,c,d | K | 4 to 8 | F |
| Slope area | ISO 1070 | M,L | M,L | M,L | | b,c,d | K,N | 10 to 50 | Q |
| Dilution, constant rate | ISO 9555 | S,M | S,M | S,M | | c,g,k | K,N | 4 to 8 | |
| | | | | | | | | | |
| Stage discharge | ISO 1100/2 | S,M,L | M,L | M,L | | b,c,d | G | 5 to 10 | |
| Slope stage discharge | TR 9123 | S,M | M,L | M,L | | c,d | G | 7 to 15 | |
| Acoustic | ISO 6416 | M | M,L | S,M,L | R | b,c,d | G | 4 to 8 | |
| Acoustic doppler | ISO 24154 | M | M | S,M | | b,c,d | G | | |
| Electro magnetic coil | TR 9213 | S,M | S,M | S,M | | b,d | G | 4 to 8 | T |
| Pumping stations | - | | | | | | G | 7 to 15 | |
| Flow measurement structures | various | S,M | | | | b,c,d | G | 3 to 6 | |

An explanation of the symbols is given in Table 4.21.

Table 4.21. Explanation of symbols mentioned in Table 4.20.

| Symbol | Definition |
| --- | --- |
| b | Flow should have no cross-currents |
| c | Channel should be relatively free from vegetation |
| d | Channel should be fairly straight and uniform in cross-section |
| g | Flow in the channel should be turbulent (even including a hydraulic jump) to ensure mixing |
| k | Channel should be free from recess in the banks and depressions in the bed |
| A | For velocity area method, with velocity observed at 0.6 times the depth, or with two point method, the minimum uncertainty may be up to 5% |
| B | For velocity area method, with velocity observed at surface, the minimum uncertainty may be up to 10% |
| C | Corrections may be required because of distance or air- and wet-line effects |
| D | Major error can be caused by pier effects |
| E | Major error can be due to drift, obstruction of boat and heaving action |
| F | This method is recommended for use only when the effect of the wind is small. Such conditions are likely to be so variable that no representative accuracies can be quoted, but usually the accuracy of this method is lower than conventional methods using current-meters. |
| G | Method suitable for more frequent discharge measurements |
| J | Quick method (less than 1 h) |
| K | Slow method (1 to 6 h) |
| L | Large width (more than 50 m) or high velocity (more than 3 m/s) or large depth (more than 5 m) |
| M | Medium width (between 5 and 50 m) or medium velocity (between 1 and 3 m/s) or medium depth (between 1 and 5 m) |
| N | Very slow method (more than 6 h) |
| Q | Approximate method used when velocity area method not feasible and slope can be determined with sufficient accuracy |
| R | Suspended material concentration should continue to be low to avoid too large a loss of acoustic signal; for the same reason, the flow should be free from bubbles |
| S | Narrow width (less than 5 m) or shallow depth (less than 1 m) or low velocity (less than 1 m/s) |
| T | May be used in rivers with weed growth and moving bed material |

Note: For this Section 4.12, use was made of the International Standard ISO 8363: Guidelines for the selection of methods.

## 4.13 INTERNATIONAL STANDARDS

The following International Standards on discharge measurements are available:

| Standard | Issued | Description |
| --- | --- | --- |
| ISO 748 | 1997 | Velocity area methods using current-meters or floats |
| ISO 1070 | 1992 | Slope area method |
| ISO 1088 | 2007 | Velocity area methods – Collection and processing of data for determination of uncertainties in measurement |
| ISO 1100-2 | 1998 | Determination of the stage discharge relation |
| ISO 2425 | 1999 | Measurement of flow in open channels under tidal conditions |
| ISO 2537 | 2007 | Rotating element current-meters |
| ISO 3455 | 2007 | Calibration of current-meters in straight open tanks |
| ISO 4369 | 1979 | Moving boat method |
| ISO 4375 | 2000 | Cableway system for stream gauging |
| ISO 6416 | 2004 | Measurement of discharge by the ultrasonic (acoustic) method |

| | | |
|---|---|---|
| ISO 8363 | 1997 | General guidelines for the selection of methods |
| ISO 9123 | 2001 | Stage-fall-discharge relations |
| ISO 9196 | 1992 | Liquid flow measurement under ice conditions |
| TR 9209 | 1989 | Determination of wetline correction |
| TR 9210 | 1992 | Measurement in meandering rivers and in streams with unstable boundaries |
| ISO 9213 | 2004 | Electromagnetic method using a full-channel-width coil |
| ISO 9555 | – | Tracer dilution methods for the measurement of steady flow<br>Part 1 General (1994)<br>Part 2 Radioactive tracers (1992)<br>Part 3 Chemical tracers (1992)<br>Part 4 Fluorescent tracers (1992) |
| TR 9823 | 1990 | Estimation of discharge using a restricted number of verticals |
| TR 9824 | 2007 | Measurement of free surface flow in closed conduits |
| ISO 9825 | 2005 | Measurement of discharge in large rivers and rivers in flood |
| TR 11328 | 1994 | Equipment for the measurement of discharge under ice conditions |
| TR 11332 | 1998 | Unstable channels and ephemeral streams |
| ISO 11627 | 1998 | Computing stream flow using an unsteady flow model |
| ISO 11655 | 1995 | Method of specifying performance of hydrometric equipment |
| TR 11656 | 1993 | Mixing length of a tracer |
| ISO 15768 | 2000 | Design, selection and use of electro magnetic currentmeters |
| ISO 15769 | 2000 | Guidelines for the application of acoustic velocity meters using the doppler and echo correlation methods |
| ISO 24154 | 2005 | Measuring river velocity and discharge with acoustic Doppler profilers (vessel mounted, downward looking) |
| ISO 24155 | 2007 | Hydrometric data transmission systems, specification of system requirements |
| TR 24577 | – | Non-contact methods for measurement of flow in open channels |
| TR 24578 | – | Guide to the application of Acoustic Doppler Current Profilers for measurement of discharge in open channels |
| | | |
| ISO | | International Standard, first or revised edition |
| TR | | Technical Report |
| TS | | Technical Specification |

CHAPTER 5

# Measurement of sediment transport

## 5.1 INTRODUCTION

The bottom of alluvial water courses, such as rivers and channels, is composed of granular material. These sediments can be transported by flowing water if the flow velocities are sufficiently high. Alternations in the flow velocities, either caused by nature (wet and dry seasons) or by human interference in the natural conditions (construction of structures in the river) will influence the magnitude of the sediment transport. The effect of this changeable sediment is erosion of the riverbed at some places and sedimentation at other locations.

In order to facilitate the design of civil engineering works in rivers and estuaries and to predict the morphological changes resulting from these works, many formulae have been developed, based on the results of experiments in nature and in hydraulic laboratories.

However, computed sediment transports are rather inaccurate. The reason for this is:
- The interaction between the water movement and the sediment transport formulae is very complex and therefore difficult to describe mathematically
- Sediment transport measurements are inaccurate, hence sediment transport formulae cannot be properly checked

Measurements of sediment transport are executed for the following purposes:
- For Irrigation: the design of stable channels, of sandtraps and of intake structures requires sufficient insight into the sediment transport phenomena.
- In River Engineering: knowledge of sediment transport is needed as a basis for the design of river works, e.g. cut-off of river bends or narrowing of a river stretch, as well as for navigation, flood control, water withdrawal and the design of reservoirs.

In this chapter attention is given to the following subjects:
- sediment yield (Section 5.2)
- bedforms (Section 5.3)
- classification of sediment transport (Section 5.4)

Table 5.1. Particle size classification according to British Standards.

| British Standards | |
|---|---|
| Clay | <2 μm |
| Fine silt | 2–6 μm |
| Medium silt | 6–20 μm |
| Coarse silt | 20–60 μm |
| Fine sand | 60–200 μm |
| Medium sand | 200–600 μm |
| Coarse sand | 600 μm–2 mm |
| Fine gravel | 2 mm–6 mm |
| Medium gravel | 6 mm–20 mm |
| Coarse gravel | 20 mm–60 mm |
| Cobbles | 60 mm–200 mm |

– sediment transport measurements (Section 5.5)
– bottom grab and bottom sampling (Section 5.6)
– grain-sizes (Section 5.7)
– intake structures on a meandering river (Section 5.8)
– International Standards (Section 5.9)

## 5.2 SEDIMENT YIELD

Sediments can be divided into two groups: cohesive and non-cohesive. Clays, the finest sediments, belong to the first group; sand and coarser sediments to the second. The distinction is usually made by particle size, as shown in Table 5.1.

The erosion products of a catchment area are washed over the fields and through stream channels to the river whereby they eventually leave the catchment area.

The total sediment outflow from a catchment area passing a control station at the outlet of that catchment area is called the *sediment yield*. It can either be expressed in tons per year, or in m$^3$ per square kilometre per year. The latter designation is the average denudation or degradation speed of the catchment area.

Table 5.2 shows some water and sediment characteristics of ten rivers. The figures are very approximate; different sources may give a variation of up to factor 2 in sediment yield. The denudation speed in mm/year is calculated from the figures tons/year, assuming a density of 1400 kg/m$^3$.

The denudation speed is lowest in flat, overgrown areas with temperate or cold climates, and in deserts where there is no water to transport erosion products. Table 5.2 shows that the denudation speed of the catchment area of the Rhine is in the order of one mm per thousand years. At the other end of the scale the Ganges and Hwang Ho have over one mm per

Table 5.2. Water discharge and sediment transport of ten rivers.

| River | Catchment area $10^6$ km$^2$ | Water discharge | | Sediment transport | | Sediment as ppm |
|---|---|---|---|---|---|---|
| | | m$^3$/s | mm/year | $10^6$ ton/ year | mm/ year | mg/1 |
| Congo | 3.7 | 44000 | 370 | 70 | 0.015 | 50 |
| Nile | 2.9 | 3000 | 30 | 80 | 0.015 | 630 |
| Wolga | 1.5 | 8400 | 180 | 25 | 0.010 | 100 |
| Niger | 1.1 | 5700 | 160 | 40 | 0.025 | 220 |
| Ganges | 1.0 | 14000 | 440 | 1500 | 1.000 | 3600 |
| Orinoco | 0.95 | 25000 | 830 | 90 | 0.065 | 220 |
| Mekong | 0.80 | 15000 | 590 | 80 | 0.070 | 170 |
| Hwang Ho | 0.77 | 4000 | 160 | 1900 | 1.750 | 15000 |
| Rhine | 0.36 | 2200 | 190 | 0.72 | 0.001 | 10 |
| Chao Phya | 0.16 | 960 | 190 | 11 | 0.050 | 350 |

year. The latter rivers carry the erosion products of a catchment area with a strong relief covered with fine, erodible material.

Whilst the denudation speed describes the overall erosion of the catchment area, the river engineer is usually more interested in the total amount of solids and in the sediment yield as a function of the water discharge.

Table 5.2 shows examples of these and it is evident that the Hwang Ho emerges as the muddiest river in the world in terms of total sediment yield as well as sediment per unit of water discharge. Rivers with small amounts of sediment (less than 100 ppm) are generally those in areas of temperate climates or with mild slopes.

## 5.3  BED FORMS

It is the general experience that the flow of two media alongside each other will cause waves: air-water, air-sand, water-sand, etc.

For the flow of water over a sandy bed, the following bed forms are classified:
– sub-critical flow, $Fr < 1$ (lower flow regime)
   *flat bed.* At values of the velocity about equal to the critical velocity, sediment transport without bed forms is possible.
   *ripples.* For sediment sizes $D < 600$ μm, ripples with lengths of 5–10 cm and heights of about 1 cm will develop with increasing shear stress. Ripples are quite irregular and three-dimensional of nature. Figure 5.1 shows a ripple bed.
   *dunes.* For all sediment sizes and increasing shear stress, dunes are developed. Dunes have a more two-dimensional character and are longer and higher than ripples.
– critical and supercritical flow (upper flow regime, $Fr \geq 1$)
   *plane bed, washed out dunes.* If the velocity further increases, the dunes will be washed away; the bed becomes flat again. Sediment transport rates are high.

Figure 5.1. Ripples.

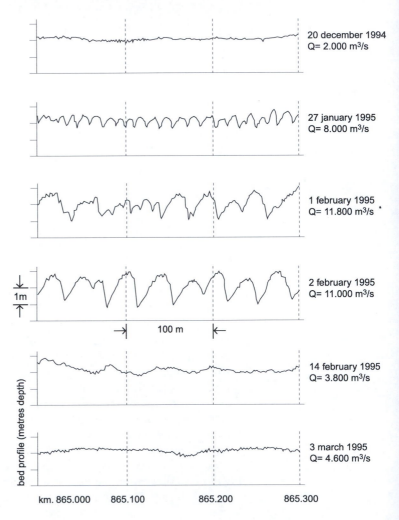

Figure 5.2. Changes in bed form during a flood on the River Rhine.

*antidunes.* Further increasing velocity gives a bed form called anti-dunes. They travel in upstream direction. The water surface is unstable and in phase with the bed forms.
*chutes and pools.* At still higher velocities chutes and pools are formed.

For fine sands the transition forms between lower regime and upper regime takes places at Froude values $Fr < 1$.
Ripples and dunes both travel in downstream direction.

During floods the bed forms will change.
Figure 5.2 illustrates different bed forms in the River Rhine at Lobith (Germany – the Netherlands border) during the floods of January 1995. The top of the floodwave occurred on 31 January 1995 with a discharge $Q = 12.100$ m³/s (once per 80 years) at a water level NAP +16.68 m. The bed forms have been measured using an echosounder, taking longitudinal profiles.

## 5.4 CLASSIFICATION OF SEDIMENT TRANSPORT

### 5.4.1 *Introduction*

The sediment transport can be classified according to origin and mechanism as illustrated in Figure 5.3.

*Definitions:*
– bed material transport: transport of material which is found in that part of the riverbed that is affected by transport
– wash load: transport of material which is not found in the riverbed, and which is permanently in suspension
– bed load: transport in almost continuous contact with the riverbed, carried forward by rolling, sliding or jumping
– suspended load: sediment, in suspension by turbulence forces, and not in contact with the riverbed

According to the mechanism of suspension the suspended sediment may belong to the bed material load and the wash load. Wash load is material finer than the bed material. It has no relation to the transporting capacity of the stream: the rate is determined by the amount which becomes available by erosion in the catchment area upstream. Usually a mean grain size $D_{50} = 60$ μm is taken as a practical distinction between wash load and bed material load. For most rivers sediment of this size (and smaller) is uniformly distributed over the vertical. The distinction between bed load and suspended load cannot be defined sharply. Not only the grain size but also the flow conditions characterize the distinction.

In some cases a free exchange of the bed material load with the bed is not possible, for instance, due to the presence of gravel layers. In that

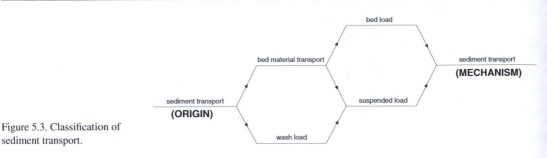

Figure 5.3. Classification of
sediment transport.

case the sediment transport may be smaller than the transport capacity of the flow: the sediment transport then depends only on the sediment supply to the reach. For an alluvial channel however, the sediment transport is generally equal to the transport capacity of the flow, with a few exceptions if the flow conditions show relatively abrupt changes in space or time.

In dynamic braided sections this is only true for mean values, averaged over time and space. Strong local temporary deviations occur in erosion and sedimentation zones.

Sediment concentrations are usually expressed in parts per million (ppm), i.e. ratio of dry weight to total weight of the water sediment mixture.

Sediment transport is generally expressed as dry weight or volume per unit of time, or as bulk volume including pores in the settled volume per unit of time.

For measurement of the three different types of sediment transport – bed load, suspended load and wash load – special instruments and methods are required. Before describing the sampling methods and the elaboration of data, the differences between types of sediment transport are mentioned.

### 5.4.2 *Bed load*

Bed load is the transport of sediment particles sliding, rolling or jumping over and near the riverbed, generally in the form of moving bed forms such as dunes and ripples (however bed forms partly move by suspended load as well). Many formulae have been developed to describe the mechanism of the bed load, some being completely experimentally founded, while others are based on a model of the transport mechanism. Most of these equations, however, have in common that they contain a number of 'constants' which have to be modified according to the field data collected for a certain river. In fact, all the deviations in bed load from the theoretical results are counteracted by selecting the right 'constants'.

Most of the available bed load functions can be written as a relation between the transport parameter

$$X = \frac{T}{\sqrt{\Delta g D^3}} \text{ (dimensionless)}$$

and the flow parameter

$$Y = \frac{\Delta D}{\mu h S} \text{ (dimensionless)}$$

where
$T$ = transport of solids in volume per unit width (often also use is made of the transport including voids $(T_\varepsilon)$; then $T_\varepsilon (1 - \varepsilon) = T$, in which $\varepsilon$ is the porosity, m³/s/m¹)
$h$ = depth of flow (m)
$D$ = grain diameter (m)
$\Delta$ = relative density = $(\rho_s - \rho_w)/\rho_w (-)$
$S$ = slope (–)
$\mu$ = so-called ripple factor, in reality a factor of ignorance, used to obtain agreement between measured and computed values of $T$. (–)

As an example of such an 'X versus Y relation', the well known Meyer-Peter/Müller bed load function may be given:

$$X = 13.3(Y^{-1} - 0.047)^{\frac{3}{2}} \tag{5.1}$$

When bed load measurements are carried out, it is important to realise that this transport takes place as the propagation of bed forms; the transport intensity on the top of the dunes is large and in the troughs small or nil.

The integration time for a bed load sampler is relatively small (2 minutes) for technical reasons, whereas the period of the fluctuations in the bed load transport varies between several hours and even days. Consequently, an estimate of the average bed load in a cross section can only be obtained by taking a large series of measurements. Measurements should therefore cover at least the time required for several dunes, to pass the control section.

### 5.4.3 *Suspended load*

Suspended load is the transport of bed particles which are in suspension when the gravity force is counterbalanced by upward forces due to the turbulence of the flowing water. This means that the particles make larger or smaller jumps, but eventually return to the riverbed. By that time, however, other particles from the bed will be in suspension and, consequently, the concentration of particles transported as suspended load does not change rapidly in the various layers.

A strict division between bed load and suspended load is not possible; in fact, the mechanisms are related. It is therefore not surprising that the so-called 'total load equations' have a similar construction as the bed load equations. Bed load and suspended load together are often called bed material load or total load (wash load is not included).

Many total load equations have been developed, such as the equation of Engelund and Hansen. These equations do not give information on the distribution of the concentration of particles in the vertical. The value of the concentration ($C$) is often determined theoretically. In most cases it is recommended to carry out suspended-load measurements, taking measuring points at various heights in each vertical in order to know the concentration-distribution in the vertical.

Figure 5.4 shows the distribution in the vertical for the various kinds of transport.

### 5.4.4 *Wash load*

Wash load is the transport of small particles finer than the bulk of the bed material and rarely found in the bed. Transport quantities found from bed load, suspended-load and total load formulae do not include wash load quantities.

However, in dynamic braided sections big quantities of silt may accumulate behind bars, in abandoned scour holes, etc., complicating the conditions considerably.

Whereas for a certain cross section quantities of suspended load and bed load can be calculated with the use of the locally valid hydraulic conditions, this is not the case for wash load. The rate of wash load is mainly determined by climatological characteristics and the erosion features of the whole catchment area.

As there is normally no interchange with bed particles, wash load is not important for local scour. Due to the very low fall velocity of the wash load particles, wash load only contributes to sedimentation in areas with low flow velocities (harbours, reservoirs, dead river branches, etc.) Due to the small fall velocity, in turbulent water the concentration of the particles over a vertical (generally expressed in parts per million, p.p.m) is rather uniform, so that even with one water sample a fairly good impression can

Figure 5.4. Distribution of concentration in the vertical (after: Hayes, 1978).

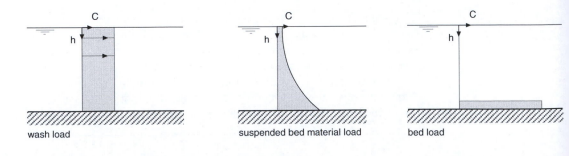

wash load          suspended bed material load          bed load

be obtained. The wash load concentration over the width, however, may vary considerably.

## 5.5 SEDIMENT TRANSPORT MEASUREMENTS

### 5.5.1 *Introduction*

The instruments and methods, mentioned in this section, were developed for the measurement of bed load, suspended-load and wash load. Many instruments – not mentioned here – are used in various parts of the world.

The great variety of instruments and methods is mainly related to the high inaccuracy which makes a clear selection of instruments and methods rather difficult.

A number of national institutes with a wide experience in hydrometric surveys are still continuing to improve existing instruments and to develop new types and methods.

The instruments described here are an arbitrary collection, subdivided according to the type of sediment load, as follows.

A more up to date overview of Sediment-Sampling Equipment and Sediment-Sampling Techniques is given in "Field Methods for Measurement of Fluvial Sediment" 2005, issued By U.S. Department of the Interior, U.S. Geological Survey (see References). The "Sediment-Sampling Equipment" section encompasses discussions of characteristics and limitations of various models of depth- and point-integrating samplers, single-stage samplers, bed-material samplers, bedload samplers, automatic pumping samplers, and support equipment. The "Sediment-Sampling Techniques" section includes discussions of representative sampling criteria, characteristics of sampling sites, equipment selection relative to the sampling conditions and needs, depth and point-integration techniques, surface and dip sampling, determination of transit rates, sampling programs and related data, cold-weather

Table 5.3. Instruments and methods for the measurement of sediment transport.

| Type of load | Instrument/method | Described |
|---|---|---|
| Bed load | – Trap samplers | |
| |    – Bed load Transport Meter Arnhem (BTMA) | 5.5.2 |
| |    – Helly Smith (HS) | |
| |    – Delft Nile Sampler (also suspended load) | |
| | – Dune tracking method | 5.5.3 |
| Suspended load | – Bottle and trap samplers | |
| |    – Delft Bottle Sampler | 5.5.4 |
| |    – US-49 Depth – Integrating Sampler | |
| | – Pump samplers | |
| | – Optical and acoustical samplers | 5.5.5 |
| Wash load | – Water sampler (bottle) | 5.5.6 |
| | – Turbidity meter | 5.5.6 |

sampling, bed-material and bedload sampling, measuring total sediment discharge, and measuring reservoir sedimentation rates.

For a more detailed description of a number of instruments of the various types, reference is made to literature (Van Rijn, 1986).

### 5.5.2 *Bed load Transport Meter Arnhem*

The Bed load Transport Meter 'Arnhem' (BTMA) is an instrument to measure the bed load of coarse sand and fine gravel just above the river bottom.

*Principle*: A frame mounted sampler is pressed on to the riverbed by a leaf spring. Behind the mouth of the sampler is a fine meshed wire basket. The shape of the basket causes a low pressure behind the instrument in such a way that water and the transported bed material enter the mouth with the same velocity as that of the undisturbed flow. The bed load particles which are too coarse to pass the meshing are caught. The BTMA catches material coarser than 300 μm (theoretical value of the meshes). Bed material between 60 and 300 μm is lost, which may be acceptable for some rivers (e.g. the Rhine), but not in general.

*Advantage:* The instrument is of simple and sturdy construction, and can easily be repaired and maintained in the field.

*Disadvantage:* Because of its weight and dimensions, a davit and winch are required for handling it. The flow velocity range in which it can be used is limited to v = 2.5 m/s due to the construction of the basket.

Measuring bed load with the BTMA or the HS, it is assumed that:
- the height of the sampler mouth corresponds with the thickness of the bedlayer
- no suspended load is entering
- the 60–300 μm fraction is negligible

*Selection of site and positioning of the BTMA*
The main requirements for the selection of the measurement site are:
- a stable river reach has to be selected in order to avoid non steady bed conditions during the measurements, and
- reliable measurements of the hydraulic conditions (depth, flow velocity, grain size and energy slope) have to be possible.

Bed load transport $s_b$ is measured in a number of verticals in a cross section. In each of these verticals a good estimate of $s_b$ is necessary. It should be that $s_b$ has a fluctuating magnitude; the 'periods' present in these fluctuations are governed by the wave period of the bed form (ripples and dunes). Except in rare cases, where dune lengths are large in comparison with the depth of water, it is not possible to place the sampler with sufficient accuracy in a particular location on a sand dune. Therefore, random sampling has to be made. Lowering the sampler from an anchored survey boat implies that measurements are carried out at

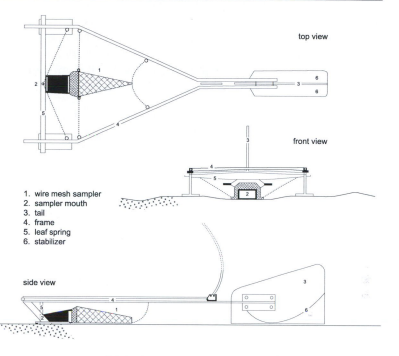

top view

front view

1. wire mesh sampler
2. sampler mouth
3. tail
4. frame
5. leaf spring
6. stabilizer

side view

Figure 5.5. Bed load Transport
Meter Arnhem (after: Nedeco,
1973).

random over a length $L$ (see Figure 5.6). This length $L$ depends on depth,
flow velocity and type of instrument and has to be relatively large in
comparison with the dominant dune length $\lambda$. If the condition $L \geq \lambda$ is
fulfilled (case A), the sampler, lowered from a fixed survey boat, reaches
the bed at a random position somewhere within the interval $L$. For $L \approx \lambda$
or $L < \lambda$ (case B), the survey boat has to take different positions, in order
to achieve a random sampling.

Obviously, it is essential to know $\lambda$ under the given circumstances.
Therefore longitudinal soundings are necessary before bed load measurements are taken.

*Sampling*
Samples of the bed load can be taken with the BTMA, which is lowered
by means of the survey vessel's davit on to the riverbed. With a stopwatch the sampling time is measured, which is normally two minutes,
after which the BTMA is hoisted aboard again and the basket with the
caught bed load sample is emptied. In case of too big or too small catches

Figure 5.6. Position vessel
with respect to sand dunes
(after: Jansen, 1979).

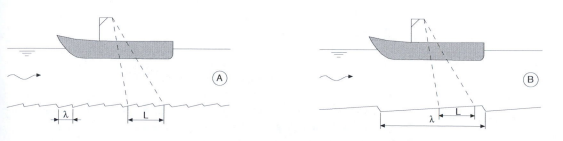

the sampling time is adjusted. The sample is measured volumetrically. Generally ten samples are taken and the results noted, averaged, and converted into a daily transport (in m³/24 hrs/m¹); see also example of elaboration form for bed load and suspended load measurements (Table 5.4).

*Elaboration*
The ten catches of the BTMA are averaged and the volume of the average catch or the complete catch is converted into daily transport (in m³/24 hrs/m¹) by means of the BTMA calibration curve (Figure 5.7). This calibration curve is based on laboratory tests.

The calibration curve of Figure 5.7 is based on the equation

$$T_i = \frac{\alpha \cdot V \cdot f}{b} \tag{5.2}$$

where
$T_i$ = bed load transport measured in a vertical (m³/24 hrs/m¹)
$\alpha$ = trap efficiency factor, based on calibration ($\alpha = 2$), not including possible losses of the 60–300 μm losses (correction to be based on sieving curves)
$V$ = catch (cm³/2 min)
$b$ = width of the mouth of the instrument ($b = 0.085$ m)
$f$ = conversion factor from cm³/2 min to m³/24 hrs ($f = 720 \cdot 10^{-6}$)

Substitution of these values in Equation 5.2 gives $T_i = 0.017\ V$, which has also been used in Table 5.4.

The total transport $T$ in the cross section becomes:

$$T = \sum b_i \cdot T_i \tag{5.3}$$

where
$T$ = bed load transport in the cross section (m³/24 hrs)
$b_i$ = part of bottom width of the river, representative for the caught $T_i$ (m)

Figure 5.7. Calibration curve BTMA (after: Nedeco, 1973).

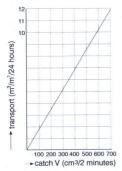

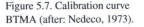

If the sediment transport has to be expressed in kg/s, the sample is taken to the laboratory and the dry weight of sediments in the sample is determined.

Now the transport in a vertical is expressed by:

$$S_i = \frac{\alpha \cdot G}{b \cdot t} \tag{5.4}$$

where
$S_i$ = bed load transport, measured in a vertical (kg/s/m¹)
$\alpha$ = efficiency factor ($\alpha = 2$)
$G$ = average dry weight of the 10 samples taken in the cross section during the time $t$ (kg)

Table 5.4. Calculation sheet bad load and suspended load measurements (after: Hayes, 1978).

RIVER: Magdalena  
STATION: Ballena  
DATE: 25-10-71 VESSEL: Explorador OBSERVERS: HJO, HG  
MEAN WATERLEVEL: —  
TIME: 9:50 TILL 10:40

| time | height above bottom | depth below surface | current velocity | sample time | caught | correction factor | caught | nozzle area | volum. trans-port | repres. height | trans-port |
|---|---|---|---|---|---|---|---|---|---|---|---|
| | m | m | m/s | min | cm³ | | cm³ | cm² | m³ | cm | m³ |
| ① | ② | ③ | ④ | ⑤ | ⑥ | ⑦ | ⑧ | ⑨ | ⑩ | ⑯ | ⑰ |
| 9:50 | | 0.70 | | 3 | 23 | 1 | 23 | 1.9 | 0.58 | 145 | 84.1 |
| | | 2.20 | | 3 | 12 | 1 | 12 | 1.9 | 0.30 | 145 | 43.5 |
| | | 3.60 | | 3 | 20 | 1 | 20 | 1.9 | 0.51 | 150 | 76.5 |
| | | 5.10 | | 3 | 30 | 1 | 30 | 1.9 | 0.76 | 135 | 102.6 |
| | 0.50 | | | 3 | 36 | 1 | 36 | 1.9 | 0.91 | 10 | 9.1 |
| | 0.40 | | | 3 | 76 | 1 | 76 | 1.9 | 1.92 | 10 | 19.2 |
| | 0.30 | | | 3 | 14 | 1 | 14 | 1.9 | 0.35 | 10 | 3.5 |
| | 0.20 | | | 3 | 53 | 1 | 53 | 1.9 | 1.34 | 10 | 13.4 |
| | 0.10 | | | 3 | 128 | 1 | 128 | 1.9 | 3.23 | 10 | 32.3 |

DELFT BOTTLE / TRANSPORT COMPUTATION

VERTICAL N° I  
total depth: 6.30 m

0.70  
1.45  
2.20  
2.9  
3.60  
4.40  
5.10

0.55  
0.50  
0.40  
0.30  
0.20  
0.10  
0.05  
0.4

B.T.M.A.

| no. | sample time | caught | caught in 2 min. | trans-port |
|---|---|---|---|---|
| | min. | cm³ | cm³ | m³ |
| ⑪ | ⑫ | ⑬ | ⑭ | ⑮ |
| 1 | | | 28 | 0.48 |
| 2 | | | 22 | 0.37 |
| 3 | | | 25 | 0.42 |
| 4 | | | 25 | 0.43 |
| 5 | | | 20 | 0.34 |
| 6 | | | 30 | 0.51 |
| 7 | | | 20 | 0.34 |
| 8 | | | 25 | 0.42 |
| 9 | | | 15 | 0.25 |
| 10 | | | 18 | 0.31 |
| | | | mean: | 0.4 |

TRANSPORT m³/m¹/24 hrs : $385$  
IN VERTICAL

$$⑩ = \frac{⑧}{⑨ \times ⑤} \times 0.144 \ \text{m}^3/24 \ \text{hours}/\text{m}^1$$

$$⑭ = \frac{⑬ \times 2}{⑫}$$

$$⑮ = 0.017 \times ⑭$$

$$⑰ = ⑯ \times ⑩$$

$b$ = width of the basket ($b$ = 0.085 m)
$t$ = sampling time (s)

The total transport $S$ in the cross section becomes:

$$S = \sum b_i \cdot S_i \qquad (5.5)$$

Normally in the laboratory also the grain size distribution of the BTMA samples will be determined (see Section 5.7).

*Other bed load samplers* (see Van Rijn 1986)
– Delft Nile Sampler: total load sampler
  The 60 kg heavy instrument is equipped with one nozzle mouth to sample bed load and seven intake nozzles operated by pumps, to sample suspended load.
– Helly Smith: modified version of the BTMA (Figure 5.8)

### 5.5.3 *Dune tracking method*

Dune tracking or bed form tracking is the determination of bed load transport from bed profiles, measured at successive time intervals, assuming steady flow conditions and a more or less two dimensional dune profile.

Figure 5.9 shows the bed profile for two successive days (interval 24 hours). The bed profile measurements are made using an analogue echosounder in a longitudinal section between two well defined transit lines, indicated by marks on both banks.

The average migration velocity $v_d = \ell_d/t$, the average dune height is $h_d$. The bed load transport in bulk volume is as follows:

Figure 5.8. Helly Smith bed load sampler.

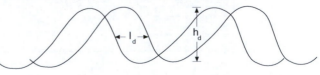

Figure 5.9. Bed profiles for two successive days.

$$T_b = b \cdot h_d \cdot \ell_d \cdot t^{-1}$$

(5.6)

where

$T_b$ = bed load transport (m³/24 hours)
$b$  = bottom width (m)
$t$  = time interval (days)
    The expected uncertainty of the method is about 40 to 50% (Van Rijn, 1986)
    Sources of uncertainty are:

– which part of the bedform propagation is caused by suspended load
– the three dimensionality of bedforms

### 5.5.4 *Delft Bottle*

*Positioning*

The Delft Bottle (Delftse Fles, D.F.) is an instrument to measure suspended load in rivers. It measures from the surface down to 0.5 m above the riverbed when suspended by a wire, and from 0.5 to 0.05 m above the riverbed when fixed in a frame (sledge). In the frame the D.F. is tilted, and the use of bent nozzles is required for the depths of 10, 20 and 30 cms above the river bottom.

*Principle:* The sediment containing water flows through a bottle-shaped sampler, the shape of which induces a low pressure at the rear end (water outlet) in such a way that the water enters the mouth of the sampler with almost the same velocity as the undisturbed flow. The inside shape of the sampler and the sharp decrease of the velocity in the wide sampling chambers causes the sediment material to settle there. This settled material can be taken out and measured volumetrically after the D.F. is out of the water.

It is possible with the D.F. to measure sediment transport with velocities up to 2.5 m/s, although the correction factor increases considerably for such high velocities. The average grain size of the sediment must exceed 0.05 mm (50 µm).

The D.F. is loosing part of the sample:
– 100% of the grains $D < 50$ µm
– partly: the grains $50 < D < 100$ µm

Therefore the sampling efficiency of the Delft Bottle is a function of the grain size distribution of the suspended material.

*Advantages:* Because of the flow through principle, a large volume of water is sampled, and it is a direct transport measurement. The D.F. is of simple and sturdy construction, can easily be maintained in the field, and can be used at any required depth.

*Disadvantage:* Because of its weight a davit and winch are required for handling.

*Sampling*

Samples of the suspended load are taken with the D.F., which is lowered into the river to the required depth by means of the davit aboard the survey vessel. The depth of the instrument is determined by the quantity of paid-out cable and indicated on a counter block. The counter block, through which the suspension cable of the D.F. runs, is put on zero when the D.F. is exactly on the water level. As soon as the D.F. is fully submerged it is kept there for a while and the instrument will incline backwards due to the air contents. The air will escape from the nozzle and a small opening at the top of the rear end. As soon as the instrument is filled with water, it is lowered quickly to the required depth. Sampling time then starts and is measured by a stop-watch. A sampling time of three minutes has proved to be sufficient. (It is recommended that the hoisting of the D.F. sampler be included in the total measuring time of the suspended load measurements). The D.F. is then hoisted aboard again, and the contents of the sampling chambers emptied into the special D.F. glass and measured volumetrically. Generally samples are taken every

Figure 5.10. The Delft Bottle (D.F.) (after: Nedeco, 1973).

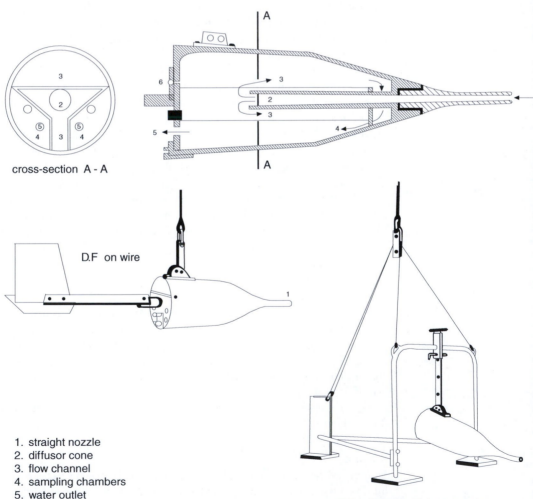

cross-section  A - A

D.F  on wire

1. straight nozzle
2. diffusor cone
3. flow channel
4. sampling chambers
5. water outlet
6. hinge

D.F. in frame, with bent nozzle

1.5 meter in the measuring vertical, as well as five samples at every
10 centimetres in the half metre just above the riverbed.

*Elaboration*
The catches with the D.F. are volumetrically measured, and noted in
cubic centimetres. The samples are caught either through the small nozzle (area 1.9 cm$^2$) or through the big nozzle (area 3.8 cm$^2$).

The elaboration is carried out along the same lines as for the bed load
measurements. Table 5.4 shows the registration and elaboration of measurements in a single vertical, carried out with the BTMA and the Delft
Bottle, giving the transport in volumes (m$^3$/day/m$^1$) for the particular
vertical.

*Other suspended load samplers (see Van Rijn 1986)*
– US-49 Depth Integrating sampler
  The sampler is lowered and raised at a uniform rate, taking samples
  throughout the time of submergence, so measuring the depth-averaged
  concentration of sediments (sand and silt particles);
– pump samplers: to obtain a reliable average sediment concentration,
  sampling period about 5 minutes;
– optical and acoustic samplers.

### 5.5.5  Optical and acoustic sampling methods

Optical and acoustical suspended load sampling methods enable the
continuous and contactless measurement of sediment concentrations,
which is an important advantage compared to the mechanical sampling
methods.

Although based on different physical phenomena, optical and acoustical sampling methods are very similar in a macroscopic sense.
In practice, the optical and acoustical sampling methods can only be used
in combination with a mechanical suspended load sampling method to
collect water sediment samples for calibration.

*Measuring range*
For an optimal sampling resolution the wave length and particle size
must be of the same order of magnitude. Therefore the optical method is
most suitable for silt particles (<50 μm). The upper concentration limit
for optical samplers is about 25.000 mg/l. The acoustic method is most
suitable for sand particles (≥50 mm). The upper concentration limit is
about 10.000 mg/l.

*Advantages*
An important advantage of optical and acoustical samplers is the continuous measurement of the suspended sediment concentration.
Further information on optical and acoustical sediment sampling methods and instruments is given in (Van Rijn, 1986): six optical techniques
and two acoustical techniques.

### 5.5.6 *Water sampler*

*Positioning*

The water sampler used to measure wash load concentrations is a heavy device in which a bottle, closed by a rubber stopper, can be placed. The sampler is suspended by a line, then lowered to the required depth, and the rubber stopper pulled off by means of the thin line fixed to it. After sufficient time has elapsed for the bottle to fill, the instrument is hoisted and the bottle taken out, closed and labelled.

*Disadvantage:* The water sampler disturbs the flow pattern and consequently cannot be used to measure the total sediment load transported by the river. The wash load, however, consisting of very fine particles, is less affected by the distortion of the flow and may accurately be estimated by the elaboration of water samples collected in this way.

Another type of water sampler (based on the same principle) is a bottle shaped perspex body of little weight, and with an opening on top which could be closed and opened by the above mentioned rubber stopper on a line. With a locally acquired weight connected to its bottom, the instrument could be used in the same way as the other sampler.

*Advantage:* The light weight of this instrument makes it possible to carry it around in hand luggage, and to use it in preliminary investigations.

*Sampling*

Samples of the wash load are taken with the water sampler, which is lowered into the river to the required depth by a hand line. The rubber stopper is then pulled off the bottle by means of a thin line allowing the bottle to be filled up with river water. Care should be taken not to hoist

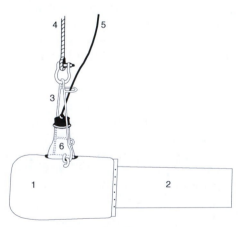

1. heavy weight metal body
2. tail, against knotting of the suspension-line and the stropper-line in currents
3. suspension bracket, also clamping the bottle
4. suspension-line
5. stropper-line
6. milk bottle (1/2 l.)

Figure 5.11. Metal Water Sampler (after: Nedeco, 1973).

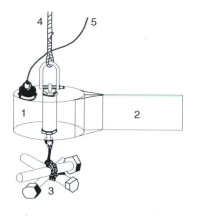

1. transparent perspex body
2. tail, against knotting of the suspension-line
   and the stopper-line in currents
3. improvised weight
4. suspension-line
5. stopper-line

Figure 5.12. Perspex Water
Sampler (after: Nedeco, 1973).

the sampler too soon, but to allow time for the bottle to be completely filled in order to prevent exchange of water content. The bottle is then taken out of the sampler, corked and labelled.

*Elaboration*
The water samples taken with the sampler and stored in bottles, have to be elaborated in the laboratory. The amount of wash load can be determined by means of the turbidity meter, or by filtering, drying and weighing of the sample. In the latter method the volume of the sample is first measured, whereafter it is filtered through a paper filter of known (dried) weight (preferably by means of a vacuum pump). The catch (wash load) is then dried for several hours in a stove, and weighed on an electronic balance with an accuracy of about 0.1 milligram. The weight, minus the weight of the paper filter, indicates the amount of wash load per volume of the sample, and thus, with known discharge of the river, the total wash load transport.

The identification of each sample is very important and should at least contain the following essential information: site, date, time, section number, vertical number, bottle number, temperature, initials observer.

The data about the sediment content are expressed in units of concentration

$$\text{so in mg/l or ppm} = \left( \frac{\text{weight dry sediment in kg} * 10^{-6}}{\text{weight water sediment mixture in kg}} \right)$$

The wash load contains the smallest grains of the grain size distribution of the total sediment transport. This portion is flowing through the Delft Bottle, because this instrument catches only particles >50 μm.

For this reason the concentration of the wash load should be sampled with a water sampler or other types of instruments that will catch material <50 µm (e.g. the US Depth integrating sediment sampler).

In general the wash load can be assumed to be evenly distributed over the vertical. Wash load concentration may vary considerably over the width of the river.

*Turbidity meter*

The turbidity meter is an optical instrument to measure the concentration of fine suspended sediment either in water samples or directly in rivers, according to the principle of determination of light absorption by a sample of river water compared with the absorption of a clear water sample. The latter should preferably be of the same river water, but cleared by filtering the silt out of it (the silt then being dried and weighted). A calibration curve must be made by plotting the concentrations found by filtering and weighing of the samples versus the extinction figures of the samples found by the turbidity meter.

## 5.6 BOTTOM GRAB AND BOTTOM SAMPLING

Although not belonging to the range of instruments that measure the bed material load transported in rivers, this device is useful to determine the material of which the riverbed is composed (top layer).

The sampler consists of a grab which is lowered in an open position by a line. When contacting the bed, the lever that keeps the grab in an open position disconnects and while hoisting, the grab is closed and holds the bottom sample.

*Advantages:* The bottom grab is of simple and sturdy construction and can easily be maintained in the field, while no davit is required to lower or hoist the instrument.

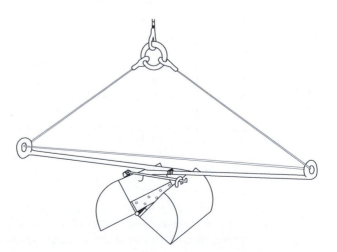

Figure 5.13. 'Van Veen' bottom grab (after: Nedeco, 1973).

*Disadvantages:* In strong flows it may be difficult to lower the sampler vertically and if it lands on the riverbed on its side, it will not grab a sample. It may then be easier to take a sample while the survey vessel is adrift.

*Sampling*

In order to gain insight into the morphology of the river, the erodibility of the riverbed and to determine the characteristics and origin of the bed material, bottom samples are taken at several locations in the river.

These locations should be chosen in such a way that a representative sampling is made across the cross section profile and along the longitu-dinal profile of the river. (Strictly, the composition should be determined for various circumstances since, for example, at high stages, layers may be uncovered which are not exposed to flow at low stages).

To describe the samples a standardized nomenclature should be used. A sand rule can be of good help. In Table 5.1 a classification based on particle sizes has been given (British Standard).

The colour of the samples should be noted on the measuring form. Caution: the colour of wet material could differ from the same dry mate-rial. If possible a description should be made of the different minerals that occur in the sample and their relative occurrence.

Also the shape of the grains should be noted.

A table in which strength and structural characteristics are given can be a good guide for field identification of bottom samples (Table 5.5). As much as possible descriptive remarks about the sample and the location should be noted down.

Even a description of the river banks may be useful for interpretation of the results or to use the bottom sample data for studying the river banks or the morphology in general.

Bottom samples can be taken with a Van Veen bottom grab.

Two types of this grab are available; small size (weight = 2.4 kg, capacity 0.5 litre), medium size (weight = 5.25 kg) and medium size + extra lead blocks (weight = 11 kg, content 2 litres).

In spite of the heavy closing force it can happen, if the grabs are sampling gravel or a mixture of sand and gravel, that a pebble sticks between the buck-ets. Be aware that in such a case the sample is not representative; the smaller parts have been lost during hoisting. It is always a good rule of thumb to take at least six samples at every location and base the conclusions on the total of all samples. This is especially important if the bottom is less regularly shaped and the bottom material consists of a mixture of materials.

For a heavy gravel bottom the 'Van Veen' grab is less useful. For this purpose a *drag grab* is better. This grab is a heavy bucket with a larger sharp circumference. It should be towed along the bottom over a distance of a few metres.

## 5.7  GRAIN SIZES

Various methods can be used for particle size analysis: sedimentation methods for particles in the clay and silt range; sieving in the case of sand and gravel; weighing when cobbles and boulders are present.

Table 5.5.  Table of strength and structural characteristics (after: Hayes, 1959).

| | | | Strength | | Structure | |
|---|---|---|---|---|---|---|
| | Types | Term | Field Test | Term | Field Identification | |
| Coarse grained, non-cohesive | Bounders Cobbles Gravel | | Can be excavated with spade, 2″ wooded peg can easily be driven in | Homogeneous | Deposit consisting essentially to one type | |
| | Uniform    Sands | Compact | Require pick for excavation, 2″ wooded peg hard to drive more than a few inches | Stratified | Alternatively layers of varying types | |
| | Graded | Slightly cemented | Visual examination. Pick remove soil in lumps which can be abraded with thumb | | | |
| Fine grained, cohesive | Low plasticity    Silts | Soft | Easily moulded in fingers. Particles mostly barely or not visible: dries moderately and can be dusted from the fingers | Homogeneant | Deposit consisting essentially of one type | |
| | | Firm | Can be moulded by strong pressure in fingers | | Alternating layers of varying types | |
| | Medium plasticity    Clays | Very soft | Exudes between fingers when squeezed in fist | Fissured | Breaks into polyhedral fragments along fissure planes | |
| | | Soft | Easily moulded in fingers | Intact | No fissures | |
| | High plasticity | Firm | Can be moulded by strong pressure in the fingers general: dry lumps can be broken, but not powdered; disintergrates under water; sticks to the fingers; dries slowly with cracks | Homogeneous stratified | Deposits consisting of essentially one type. Alternating layers of varying types if layers are thin, the soil may be described as laminated | |
| | | Stiff | Cannot be moulded in fingers | | | |
| | | Hard | Brittle or very tough | Weathered | Usually exhibits crumbs or columnar structure | |
| Organic | Peats | Firm | Fibre compressed together, colour brown to black | | | |
| | | Spongy | Very compressible and open structure, colour brown to black | | | |

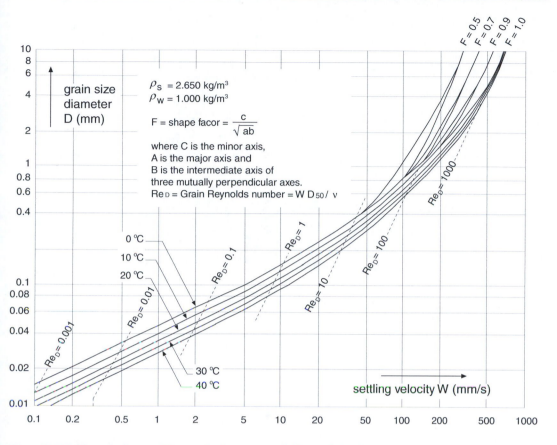

Figure 5.14. Settling velocity as a function of particle size, shape factor and grain Reynolds number.

The analysis may result in particle size distribution curves. From these curves the information needed for computation of bed material load can be read.

If many samples are involved and the only information required concerns the percentages in which some characteristic grain sizes are present in the mixture, the processing of sieve data can be programmed and the presentation of the output adapted to this purpose.

Other factors used to identify sediments are the particles' shape, density and settling velocity.

The shape of the particles is characterized by the following parameters of which only the first one has practical value:
– *shape factor*: $c/\sqrt{(ab)}$ in which $a$, $b$ and $c$ are respectively the longest, intermediate and shortest of the three mutually perpendicular axes of the particle. Most natural sand participles have a shape factor of about 0.7;
– *sphericity*: ratio of surface area of a sphere with equal volume as the particle to the surface area of the particle considered;
– *roundness*: ratio of average radius of curvature to radius of circle inscribed in the maximum projected area of the particle.

| Sample n° | | MITCH | | | D90 = 520 µm |
| date: 18 August 1972 | | | | | D65 = 340 µm |
| location:  Section Ballena | | | | | D50 = 290 µm |
| Remarks: | | ADENAVI  NEDECO | | | D35 = 270 µm |
| 440m. from the right bank | | | | | Dm = 340 µm |
| | | SIEVE ANALYSIS FORM | | | |

| Data of the lab. | | sieve curve | | | $D_m$ | | fall velocity | |
|---|---|---|---|---|---|---|---|---|
| mesh width (µm) $D_i$ | weight of material (gr.) | % on the sieve $P_i$ | \multicolumn{2}{c}{accumulated} | $\sum \dfrac{P_i \cdot D_i}{100}$ | $W_i$ (cm/s) | $\sum \dfrac{P_i \cdot W_i}{100} = \overline{W}$ | |
| | | | % on | % through | | | | |
| 4.800 | 2.4411 | 0.38 | 0.38 | 99.62 | 18.24 | | | |
| 3.400 | 1.7454 | 0.27 | 0.65 | 99.35 | 9.18 | | | |
| 2.400 | 2.1369 | 0.33 | 0.98 | 99.02 | 7.92 | | | |
| 1.700 | 1.7756 | 0.28 | 1.26 | 98.74 | 4.76 | | | |
| 1.200 | 0.8152 | 0.13 | 1.39 | 98.61 | 1.56 | | | |
| 850 | 11.2605 | 1.75 | 3.14 | 96.89 | 14.88 | | | |
| 710 | 9.3621 | 1.45 | 4.59 | 95.41 | 10.30 | | | |
| 600 | 7.2702 | 1.13 | 5.72 | 94.28 | 6.78 | | | |
| 500 | 35.9903 | 5.59 | 11.31 | 98.69 | 27.95 | | | |
| 420 | 45.1928 | 7.01 | 18.32 | 81.68 | 29.44 | | | |
| 350 | 80.9600 | 12.57 | 30.89 | 69.11 | 44.00 | | | |
| 300 | 88.9037 | 13.80 | 44,69 | 55.31 | 41.40 | | | |
| 250 | 185.7796 | 28.85 | 73.54 | 26.46 | 72.13 | | | |
| 210 | 104.2715 | 16.19 | 99.73 | 10.27 | 34.00 | | | |
| 175 | 34.9966 | 5.43 | 95.16 | 4.84 | 9.50 | | | |
| 150 | 14.7413 | 2.29 | 97.45 | 2.55 | 3.43 | | | |
| 125 | 9.6216 | 1.49 | 98.94 | 1.06 | 1.96 | | | |
| 105 | 2.8867 | 0.45 | 99.39 | 0.61 | 0.47 | | | |
| 90 | 1.9222 | 0.30 | 99.69 | 0.31 | 0.27 | | | |
| 75 | 0.9871 | 0.15 | 99.84 | 0.16 | 0.11 | | | |
| 50 | 0.6152 | 0.10 | 99.94 | 0.06 | 0.05 | | | |

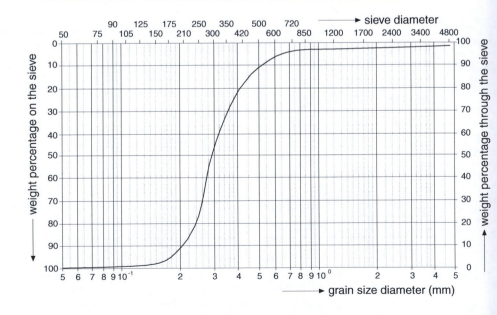

Figure 5.15. Sieve analysis form and sieve curve (after: Nedeco, 1973).

The density of a particle depends on the mineral composition. Most particles in the sand and gravel range consist of quartz which has a density of about 2650 kg/m³. For most practical computations this value can be used. However, measurement of the density might be necessary for special studies.

The settling velocity (or fall velocity) of a particle depends on its diameter and the shape factor $F$.

$$F = \frac{c}{\sqrt{ab}} \tag{5.7}$$

Figure 5.14 gives this relation for material with a density $\rho_s = 2650$ kg/m³.

*The main grain size*
The samples caught by the BTMA or bottom grab are analysed on grain size distribution by sieving. A series of sieves gradually decreasing in sieve diameter (from 4.8 mm to 0.15 mm, according to the National Standard Norm) placed on top of each other and connected to a vibrator, gives the amount of particles of each sample distributed over several sieves. The contents of each sieve are carefully measured (weighed) on an electronic balance with an accuracy of 0.1 milligram.

The sieve diameter and the weight of the caught particles, as well as their percentage of the complete sample, are used to form the sieve curve (Figure 5.15). Many natural sediments have approximately a log-normal grain size distribution, and therefore the sieve curve is normally plotted on logarithmic paper.

The mean grain size $D_m$ is determined by the relation $D_m = \Sigma^i p_i D_i$, in which $p_i$ is the percentage of a sample in the i-th sieve fraction and $D_i$ is the arithmetic mean value of the fraction limits.

## 5.8  INTAKE STRUCTURES ON A MEANDERING RIVER

In this section (after Lauterjung, 1989) an application of sediment transport in river bends is given. The location of an intake structure must be so selected that the majority of the bed load remains in the river, and is not taken in the diversion canal (suspended load must be removed by means of a sand trap).

For intake structures on a bent river section where water is not dammed up, the bed load transport is governed by spiral flow.
Figure 5.16 shows the effect of spiral flow in river bends on the bed load transport.
The spiral flow transports the bed load to the inside bend of the river.

If intake structures are designed in a bend river section, they shall be situated on an outside bend, preferable at some distance downstream of the bend where the spiral flow is strongest, as indicated in Figure 5.17.

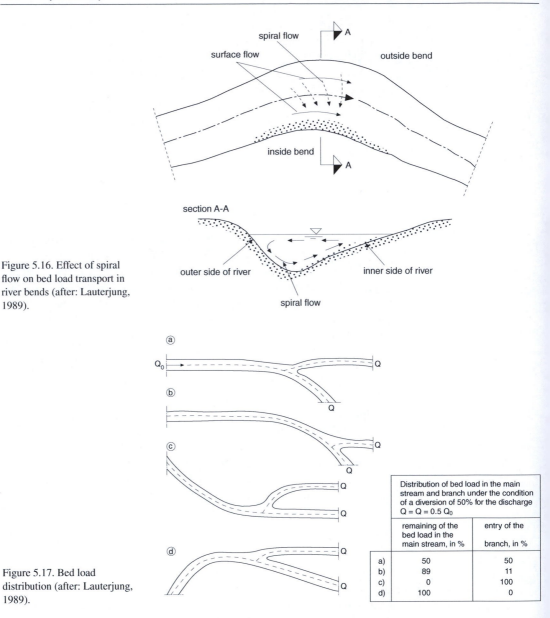

Figure 5.16. Effect of spiral flow on bed load transport in river bends (after: Lauterjung, 1989).

Figure 5.17. Bed load distribution (after: Lauterjung, 1989).

| | Distribution of bed load in the main stream and branch under the condition of a diversion of 50% for the discharge $Q = Q = 0.5 Q_0$ | |
|---|---|---|
| | remaining of the bed load in the main stream, in % | entry of the branch, in % |
| a) | 50 | 50 |
| b) | 89 | 11 |
| c) | 0 | 100 |
| d) | 100 | 0 |

Figure 5.17 shows the distribution of bed load in the main stream and the intake branch, without additional structures, and assuming a diversion of 50% for the discharge (according to Lauterjung, 1989).

## 5.9  INTERNATIONAL STANDARDS

The following International Standards are available:

ISO 3716-2006   Functional requirements and characteristics of suspended sediment load samplers.

ISO 4363-2002    Methods for measurement of characteristics of suspended sediment

ISO 4364-1997    Bed material sampling

ISO 4365-2005    Sediment in streams and canals – Determination of
concentration, particle size distribution and relative
density in reservoirs

ISO 9195-1992    Sampling and analysis of gravel-bed material

TR 9212-2006    Methods of measurement of bed load discharge

ISO 11329-2001    Measurement of suspended sediment transport in tidal
channels

CHAPTER 6

# Flow measurement structures

## 6.1 INTRODUCTION

Flow measurement structures are defined as hydraulic structures, installed
in open channels or in closed conduits with a free water level where in
most cases the discharge can be derived from the measured upstream
water level. Figure 6.1 shows a flow measurement structure.

In fact, such a structure is an artificial reduction of the cross sectional
area in the channel or pipe which causes an increase in the upstream water
level, thus creating a drop in water level over the structure. Provided the
reduction is strong enough, we have a unique relation between the discharge and the upstream water level. And by measuring this water level
continuously we can also obtain a continuous record of discharges as a
function of the time.

The relation between the discharge and the upstream water level depends
primarily on the shape and dimensions of the structure, and only slightly
on the geometry of the upstream channel or pipe. The relation can be set
up from a theoretical approach which is to be supported by a calibration,
mostly carried out by a hydraulic model study.

During the past centuries numerous types of flow measurement structures have been designed whose characteristics meet modern demands
of water resources development, particularly in irrigation schemes and
hydrological studies.

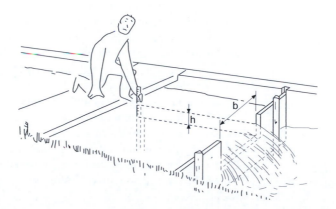

Figure 6.1. Measurement of
the upstream water level.

The most effective way to obtain a good understanding of the use of flow measurement structures is to consult a handbook (Bos 1989, Ackers 1980 and the ISO Standards) which is especially issued on these structures. Such a handbook not only gives a rather complete review of existing structures but it also provides the necessary basic principles and practical outlines how to select the most appropriate structure for specific demands and how to make the hydraulic design of a flow measurement structure.

This chapter deals with flow measurement structures in open channels like weirs, flumes and gates. In addition, some of these structures are used in closed conduits with a free water level, for instance in sewers.

## 6.2  FUNCTIONS OF STRUCTURES

In all water conservation systems natural flow through canals can be controlled by human intervention. Both the responsible authority and – in some cases – the individual farmer have tools to control the water level and the amount of flow, to answer supply and demand of water.

The hydraulic structures necessary to control level and flows are weirs, gates and flumes. The following functions can be identified:

– *upstream water level control*
  Examples are check structures and cross regulators in irrigation canals, and drop structures in steep natural streams and canals.
– *flow measurement*
  When the application is limited to flow measurement only, the structure does not have adjustable parts: the crest or sill has a fixed elevation. In general flow measurement structures are used in:
  – natural streams
  – irrigation and drainage canals
  – water purification plants and industries
  – hydraulic laboratories.
  As long as the downstream water level does not affect the flow, the discharge depends exclusively on the upstream water level. Modern flow measurement stations are equipped with micro-processors which convert the measured heads directly into digital records of discharges. If desired, discharges can be totalized to flow volumes over a certain time interval, for instance per hour or per day.
– *flow regulation and measurement*
  These structures are basically designed to regulate and to measure the flow for an almost constant or sometimes a varying upstream water level. Examples of large structures are headworks on rivers and irrigation canals. Farm turnouts can be considered as small hydraulic structures. Figure 6.2 shows an overflow structure with a movable crest, a so called Hobrad weir (*h*orizontal, *b*road-crested, *ad*justable).

Figure 6.2. The Hobrad weir.

- *flow division and measurement*

  In irrigation canals the main flow has to be distributed sometimes proportionally into two or more branches. An example to illustrate this are the division boxes in irrigation canals. These boxes are not adjustable (open or closed). In other structures the percentage distribution ratio can be adjusted by movable parts.

- *removal of excess flow*

  Part of the incoming flow both in reservoirs and in irrigation canals will not be used. This surplus water has to be drained off. Examples here are overflow structures and radial gates in diversion dams.

A clear insight into the structure's function is essential as it provides many relevant answers to questions which may arise when an appropriate type must be chosen. However, also the field conditions as well as other specific demands play an important role in this process and for this reason no simple schedule can be made up which directly relates the required function to a particular structure.

A summary of the above mentioned functions is presented in Table 6.1.

## 6.3 STRUCTURES IN IRRIGATION SCHEMES

### 6.3.1 *Introduction*

Numerous structures are necessary for the proper operation of canals. Some of them are used to measure discharges. A layout of the water distribution system is given in Figure 6.3.

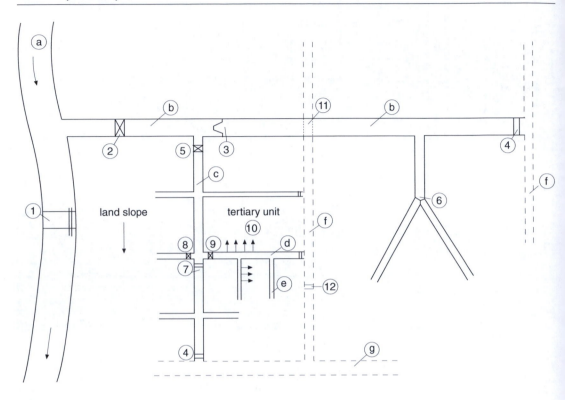

Figure 6.3. Layout of an
irrigation scheme.

The following water courses and structures are identified (in brackets
the synonymous names are given):

*Water courses*
a. Source of supply: river or reservoir
b. Main canal (primary canal) ⎫
c. Secondary canals (laterals) ⎭ feeder canals
d. Tertiary canals (sublateral, delivery canals)
e. Farm ditches
f. Secondary drain
g. Main drain

*Structures*
1. Diversion dam with spillway
2. Headwork (head regulator and intake work)
3. Cross regulator (control structure)
4. Tail structure (escape)
5. Secondary offtake (offtake structure)
6. Division structure
7. Check structure
8. Farm outlet (turnout, offtake). Individual delivery ⎫ block
9. Tertiary offtake (tertiary inlet). Group of farmers ⎭ intakes
10. Small farm intakes
11. Culvert (inverted syphon)
12. Drop structure (chute structure)

Table 6.1. Summary of the functions of measurement/regulating structures.

| Function | Structure name | Adjustable parts | Applied in |
|---|---|---|---|
| Upstream water level control | Check structures | Sometimes | Irrigation canals |
| | Cross regulators | Sometimes | |
| | Drop structures | No | Natural streams |
| | Stoplogs | Yes | All types of |
| | Flapgates | Yes | watercourses |
| Flow measurement | Many weirs and flumes | No | All types of watercourses |
| Flow regulation and measurement | Headworks | Yes | Irrigation canals |
| | Offtakes, turnouts | Yes | |
| Flow division and measurement | Division structures | Yes | Irrigation canals |
| | Division boxes | Yes | |
| Removal of excess flow | Spillways | Sometimes | Reservoirs |
| | Escapes, wasteways | Sometimes | Irrigation canals |

### 6.3.2 *Brief description of structures*

*A diversion dam* is an overflow structure built across a river to maintain the water level above the floor of the intake structure of the main canal. Diversion dams may be provided with sediment flushing facilities.

In other cases only a small part of the dam acts as an overflow structure. This is for example the case in reservoir dams, where the excess of water is removed by an emergency *spillway*.

*The headworks* (head regulator) serves to regulate the flow into the irrigation main canal. On small installations a simple slide gate may be sufficient.

For large canals flow can be controlled by one of the following structures:
– stop logs
– adjustable undershot gates
– adjustable overflow weirs

To avoid sedimentation in the irrigation system, a sand trap can be designed in the upper part of the main canal.

*Cross regulators and check structures.* Their main purpose is control of the water level at the offtake point. The water level may be kept at a nearly constant value despite variation in discharge (see Section 6.3.3).

A very small level variation at the offtake point requires a very wide overflow structure as cross regulator. When the structure needs to be wider than the canal width is, then the weir crest can be folded zig-zag. Examples are the horse shoe weir, the duck bill weir and the Giraudet weir.

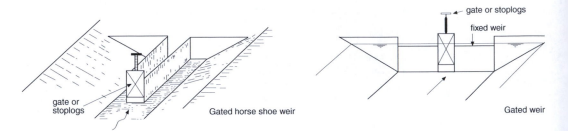

Figure 6.4. Cross regulators consisting of a fixed weir and an adjustable section.

Different types of structures can be applied as cross regulator or check:
– overflow structures with a fixed crest
– overflow structures with an adjustable crest (need of skilled personnel)
– overflow structures with shuttergates (stepwise regulation)
– overflow with stoplogs (stepwise, but difficult to operate)
– adjustable undershot gates (need frequent adjustment)
– automatic adjustable gates (a more or less constant upstream or downstream water level is maintained by float control)
Generally overflow structures are designed in the case of moderate flow variations while undershot gates are applied for large flow variations. In many situations a combination of a horse shoe weir and an adjustable structure gives a rather flexible solution:
– the horse shoe controls short term fluctuations
– the adjustable part -stoplogs or gates- regulates different flow demands.

*Tail structures and emergency structures (escapes)*
The purpose is to maintain a safe maximum water level below the canal bank and to discharge water into a natural or artificial drain canal. Different situations and types are:
– overflow structures with fixed crest at the end of irrigation canals
– side weirs which are also overflow structures with a wide crest, but situated in one of the banks
– emergency syphons (need often an energy dissipator)
– emergency automatic adjustable gates
– emergency automatic overflow flap gates
– vertical pipes in canals and reservoirs for security overflow (need energy dissipator)

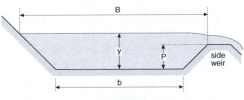

Figure 6.5. Side weir (after: Bos, 1989).

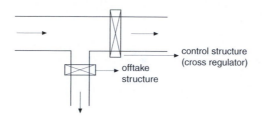

control structure
(cross regulator)

offtake
structure

Figure 6.6. Offtake.

Hobrad weir

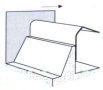

Butcher weir

Figure 6.7. Adjustable
overflow structures (after: Bos,
1989).

Radial gate

Constant head orifice

Figure 6.8. Adjustable orifices
(after: Bos, 1989).

An energy dissipator is a stilling basin in which the energy, caused by
high velocities, is dissipated locally, so preventing erosion downstream
of a structure.

Tail structures need a provision to empty the canal for the purpose of
maintenance of canals and structures.

*Offtake structures*

Purpose: control of flow into secondary and tertiary canals.

Essentially secondary offtakes do not differ from tertiary offtakes.

The dimensions depend on the maximum discharge. For tertiary
offtakes the discharge ranges from $Q = 20$ ℓ/s to $Q = 150$ ℓ/s. For secondary offtakes the design discharge is an multiple of this.

Various types of structures are applied as offtake:

– flumes ensure a proportional distribution. Discharges depend on the
  water level in the feeder canal, and are not adjustable. Flumes are suitable in canals carrying a high amount of silt.

Examples of adjustable overflow structures are:

– Romijn weir (Indonesia)
– Round nose horizontal broad crested weir (Hobrad weir)
  This structure is an improvement of the Romijn weir. Mechanically
  the structures operate in the same way.
– Butcher's weir (Sudan)
– Shutter gates (stepwise adjustable)
– Stoplogs (stepwise adjustable)

Examples of undershot gates which need small head losses are:

– Neyrpic modules with one or more openings of different width.
  The structure gives a nearly constant discharge, independent of the
  upstream water level. Discharges are stepwise adjustable by closing
  one or more openings.
– adjustable orifices with concrete blocks (India, Pakistan)
– Crump-de Gruyter adjustable streamlined gate (Indonesia)
– radial or tainter gates
– Constant Head Orifices (USA), which are hard to operate

Examples of pipe flow are:

– Metergate USA (adjustable)
– pipe outlets (not adjustable)

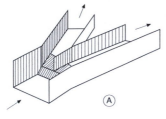
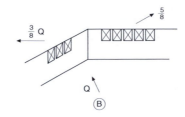
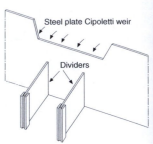

Figure 6.9. Examples of proportional divisors.

*Division structures and boxes*
Two types of divisions can be identified

1. Division structures:
   – proportional divisors as overflow structure with a fixed crest. The division ratio is constant:
   – adjustable divisors using shutter gates, sliding gates or automatic gates. The division ratio may be varied.
2. Division boxes are applied where small channels (offtakes) are branching off from larger ones.
   In many cases water distribution especially for the tertiary units takes place by the use of division boxes, which are alternative solutions for the combination of two or more tertiary offtakes and a check structure (Figure 6.10).
   Flow distribution in boxes can be realized in different ways:
   – proportional division by flumes or fixed weirs
   – adjustable division by undershot gates
   – adjustable division by stoplogs or adjustable overflow structures

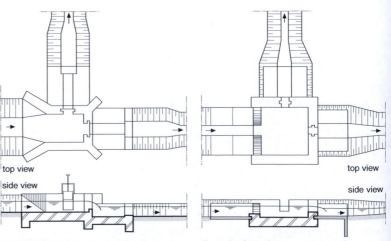

Figure 6.10. Tertiary boxes.

Example of a tertiary or quarternary box          Example of a tertiary or quarternary box

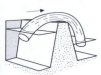

Siphon

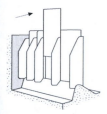

Neyrpic modules

Figure 6.11. Small farm
intakes (after: Bos, 1989).

*Small farm intakes*

Purpose:  supply of water from a tertiary canal or from farm ditches
to the field. The structures either release the full flow or are
not in use.

The maximum discharge: $Q = 40$ ℓ/s.

Different types are:
- syphons and small pipes $d \leq 0.07$ m.
  The flow is modular and heads vary from 0.10 m to 0.20 m.
  Supply is stopped by taking out the syphon or by tapping it with a
  grass sod. Only stepwise regulation is possible by taking out part of
  the total number of syphons.
- pipes $d > 0.20$ m.
  The flow may be either free or submerged.
  Large flows will occur when the (differential) head is 0.20 m or more,
  while the velocity $v \geq 2$ m/s may cause erosion at the downstream
  side. Use of this structure should not be recommended, unless they are
  provided with commercial gates.
- Neyrpic modules (see offtake structures).

*Drop structures*

The main purpose is to avoid canal erosion in steep sloping canals.
Especially natural drains need to be protected against a continuous degradation of their bed.

Two different types are feasible:
1. straight drop structures. They consist of a weir with vertical back face
   and stilling basin to dissipate the energy.
   Examples are:
   - weir sill
   - several short crested weirs with fixed crest and flap gates.

2. inclined drops or chutes, having a sloping downstream face and also
   provided with a stilling basin.

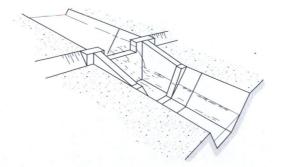

Figure 6.12. Straight drop
structure.

Table 6.2. Irrigation structures.

| Name of structure | overflow $a_1$ / underflow $a_2$ / modular $b_1$ / submerged $b_2$ | Head-works | Cross regulators Q = $Q_{FS}$ check structures | Cross regulators Q < Q < $Q_{FS}$ check structures | Tail and emergency structures | Secondary offtakes | Tertiary offtakes | Small farm intakes | Division structures | Drop structures |
|---|---|---|---|---|---|---|---|---|---|---|
| Adjustable overflow weirs | $a_1$ $b_1$ | x | | x | | | | | | |
| Stoplogs | $a_1$ $b_1$ | x | | x | | | x | | | |
| Adjustable undershot gates | $a_2$ $b_2$ | x | | x | | x | x | | x | |
| Overflow weirs with fixed crest | $a_1$ $b_1$ | | x | | x | | | | | x |
| Horse shoe/duck bill/Giraudet | $a_1$ $b_1$ | | x | | x | | | | | x |
| Flash board checks | $a_1$ $b_1$ | | | x | | | | | | |
| Combination horse shoe + gate | $a_{1,2}$ $b_{1,2}$ | | | x | | | | | | |
| Shutter gates | $a_1$ $b_1$ | | | x | | x | | | x | |
| Automatic gates | $a_{1,2}$ $b_1$ | | | x | x | | | | | |
| Syphon spillways | $a_2$ $b_1$ | | | | x | | | | | |
| Side weirs | $a_1$ $b_1$ | | | | x | | | | | |
| Romijn and Butcher weir | $a_1$ $b_1$ | | | | | x | x | | | |
| Constant Head Orifice | $a_2$ $b_2$ | | | | | x | x | | | |
| Crump-de Gruyter adj. orifice | $a_2$ $b_1$ | x | | | | x | x | | | |
| Crump overflow weir | $a_1$ $b_1$ | | | | | x | x | | | |
| Parshall flume | $a_1$ $b_1$ | | | | | x | x | | | |
| Metergates | $a_2$ $b_2$ | | | | | x | x | x | | |
| Sharp-crested weirs | $a_1$ $b_1$ | | | | | x | x | x | | |
| Flumes in proportional divisors | $a_1$ $b_1$ | | | | | | x | x | x | |
| Neyrpic Modules | $a_2$ $b_1$ | | | | | | x | x | | |
| Pipe outlets | $a_2$ $b_{1,2}$ | | | | | | x | x | | |
| Syphons and small pipes | $a_2$ $b_{1,2}$ | | | | | | x | | x | |
| Adjustable divisors | $a_1$ $b_1$ | | | | | | | | | |

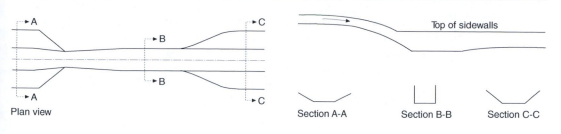

Plan view                              Section A-A      Section B-B      Section C-C

Figure 6.13. Inclined drop structure.

Essentially spillways in reservoir dams are chutes with very high drop heights.

A summary (not complete) of irrigation structures is given in Table 6.2.
Finally, the irrigation structures are classified in Section 6.5.1 to their functions.

### 6.3.3  Flow control systems

Possible flow control systems are:
– proportional: flow control with preset ratios based on command areas
– upstream control
– downstream control

In most cases the cross regulator is situated immediately downstream of the offtake, thus creating an upstream control system.
Occasionally the cross regulator is located at a short distance upstream of the offtake: this is a downstream control system.

*Comparison of downstream and upstream control.* Downstream control is suitable for on demand systems of operation. As demand for water increases, the canal control gates open so that more water passes downstream. This control system will only be successful, however, if sufficient water is available at the head of the irrigation canal system. Since some storage is available in the channel, a sudden increase in demand can usually be satisfied and off-channel storage tanks are usually not necessary. On the other hand, when water availability is limited and the volume of water that each farmer can use is specified, then an upstream control system will only allow a certain amount of water to be taken from the headworks, and each canal will carry exactly the scheduled

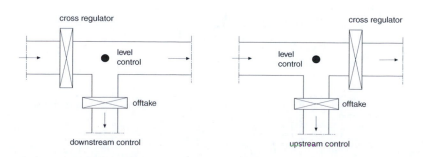

Figure 6.14. Downstream and upstream control using cross regulators.

flow. A possible penalty for using upstream control for a whole irrigation system is that irrigated lands in the most remote part of the system may frequently not receive any water – then the farmers will abandon those lands. A combination which has often been used is to have an upstream control system for the main and branch canals, and a downstream control system for the other, smaller canals.

## 6.4  CLASSIFICATION OF FLOW MEASUREMENT STRUCTURES

Discharge measurement structures are classified according to the shape of the crest in the flow direction. They can further be subdivided according to the different cross sections.

### 6.4.1  *Broad-crested weirs*

The length of the crest should be sufficient to allow straight and parallel streamlines at least along a short distance above the crest. The crest height with respect to the bottom of the approach channel must comply with a certain minimum value. The best-known structures are the following weirs:
– the round-nose horizontal broad-crested weir (Figure 6.16)
– the rectangular broad-crested weir
– the Romijn measuring and regulating weir
– the trapezoidal profile weir (Figure 6.15)
– the Fayoum standard weir
– the V-shaped broad-crested weir.

### 6.4.2  *Sharp-crested weirs*

Figure 6.15. Typical broad-crested weir with fixed crest (after: Bos, 1989).

The length of the crest is 1 to 2 mm. For this reason they are also called thin-plate weirs. The nappe is completely free from the weir body after

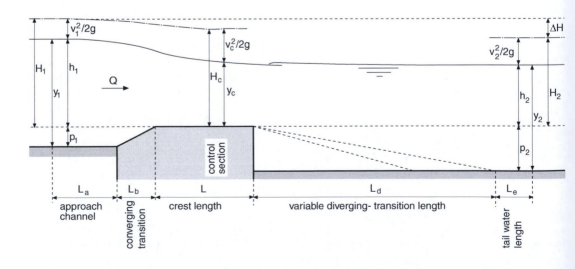

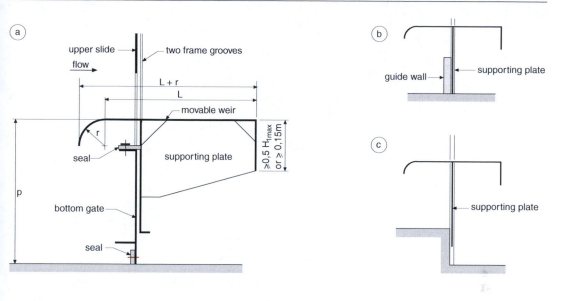

Figure 6.16. Hobrad weir (horizontal, broad-crested, adjustable) (after: ISO).

passing the weir, and the streamlines above the crest are strongly curved. In the air-filled area below the underside of the outflowing jet atmospheric pressure should prevail. Among the most used thin-plate weirs are:
– the horizontal sharp-crested weir (Rehbock, Figure 6.17)
– the rectangular sharp-crested weir (with side contraction)
– the V-shaped sharp-crested weir (Thomson)
– the trapezoidal sharp-crested weir (Cipoletti)
– the circular sharp-crested weir
– the proportional weir (Sutro weir).

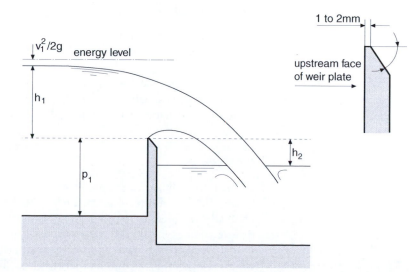

Figure 6.17. The sharp crested weir (after: Bos, 1989).

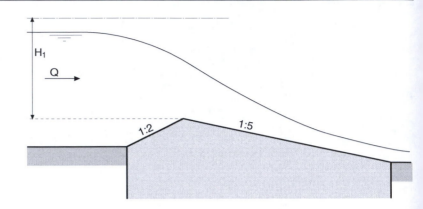

Figure 6.18. Triangular profile weir (Crump weir).

### 6.4.3 *Short-crested weirs*

We call weirs short-crested when their characteristics in some way look like those of broad-crested and sharp-crested weirs. The streamlines above the crest are curved. Well-known examples are:
– weir sill with rectangular control section
– V-notch weir sill
– the triangular profile weir (Crump weir, Figure 6.18)
– the flat V-weir
– Butcher's movable standing wave weir
– WES-standard spillways
– cylindrical crested weir (Figure 6.19)
– the streamlined triangular profile weirs
– flap gates
– the Rossum weir

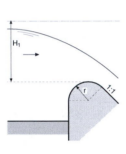

Figure 6.19. Cylindrical
crested weir.

### 6.4.4 *End depth methods*

Where the bottom of the canal drops suddenly, a free overfall is created. The water level is measured exactly above the drop (end depth or brink depth). The discharge is a function of both the end depth and the shape of the cross section. There we can identify:
– rectangular channels with a free overfall
– non-rectangular channels with a free overfall.

### 6.4.5 *Flumes*

Critical depth flumes and broad-crested weirs have some resemblance. Flumes are less restricted in crest height and the downstream section is gradually divergent to regain energy. Distinction is made between long-throated flumes and short-throated flumes. Long-throated flumes are similar to broad-crested weirs (parallel streamlines). Short-throated flumes behave like short-crested weirs (curved streamlines).

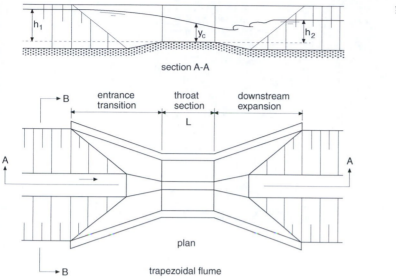

section A-A

sectional view B-B

B

entrance
transition

throat
section

downstream
expansion

L

A

A

plan

B

trapezoidal flume

Figure 6.20. Trapezoidal flume
(after: Bos, 1989).

The following long-throated flumes are mentioned:
– rectangular flumes (Venturi-flumes)
– trapezoidal flumes (Figure 6.20 and 6.21)
– U-shaped flumes.

Figure 6.21. A trapezoidal
flume in the Geleenbeek.

All other flumes are called short-throated flumes:
– throatless flume with rounded transition

Figure 6.22. A Khafagi-
venturi.

- throatless flume with broken plane transition (cut-throat flumes)
- Khafagi-venturi (Figure 6.22)
- Parshall flumes (22 different widths)
- Saniiri flumes
- H-flumes
- San Dimas flume (and modified San Dimas flume)
- Palmer Bowlus flumes (for use in conduits).

### 6.4.6 *Gates*

Each opening in a plate or a wall, the top of which is placed at a sufficient distance below the upstream water level, is a gate or an orifice.

Water flows through the opening, which is called gate flow, orifice-
flow, underflow or undershot flow. Flow may either be free or submerged.
Distinction is made between sharp-edged orifices in thin plates and a
variety of gates:

- sharp-edged orifices (rectangular, circular and other shapes)
- constant-head orifice
- radial gate (tainter gate, Figure 6.23)
- Crump-de Gruyter adjustable gate
- vertical gates (sluice gate, section 6.7, Figure 6.29)
- Neyrpic modules
- various valves
- culverts (measurement of peak flows).
- the Romijn gate
- syphons

The majority of the above mentioned flow measurement structures is
constructed between vertical side walls, thus creating two-dimensional
flow. Other structures show three-dimensional flow.

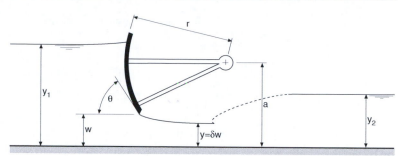

gate with sill at streambed elevation

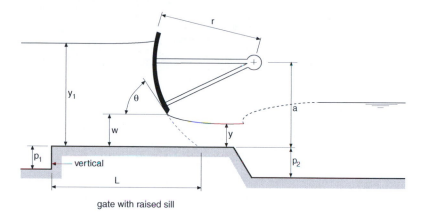

Figure 6.23. Flow below a
radial or tainter gate (after:
Bos, 1989).

gate with raised sill

For detailed information on all the different flow measurement structures, reference is made to the relevant handbooks and other literature, as
mentioned in the References.

## 6.5  FIELDS OF APPLICATION

Flow measurement structures are applied in several fields:

### 6.5.1  *Irrigation and drainage projects (agriculture)*

Flow is regulated, measured and distributed. All types of weirs, flumes
and gates are applied in irrigation schemes. Table 6.3 gives a review of
the irrigation structures and their functions.

Figure 6.24. Venturi flume and
sharp-crested weirs.

flumes                                                    sharp-crested weirs

Table 6.3. Functions of Irrigation structures.

| Structures | Function | | | |
|---|---|---|---|---|
| | Level control | Flow regulation | Flow measurement | Flow removal |
| Headworks | X | X | X | |
| Cross regulators and check structures $Q = Q_{FS}$ (FS = full supply) | X | | 0 | |
| Cross regulators and check structures $0 < Q < Q_{FS}$ | X | X | X | |
| Tail and emergency structures | X | | 0 | 0 |
| Structures in secondary and tertiary canals | X | X | X | |
| Small farm intakes | X | | X | |
| Division structures | X | X | X | |
| Drop structures | X | | 0 | |

X = main function
0 = additional function

The reader may observe that the majority of the irrigation structures have the combined function of flow regulation and flow measurement. (See also Section 6.3)

### 6.5.2 *Sanitary engineering and industry*

In most water-purification plants the effluent flow is measured, often in combination with water quality measurements.

In some cases the effluent is discharged and measured in closed conduits. In other situations the purified water leaves the plant through open channels provided with a weir or a flume.

Traditionally the selection of flow measurement structures here is rather limited and not very logical.

The following types are applied (Figure 6.24):
– sharp-crested weirs, such as the rectangular, the Cipoletti and the V-notch;
– long-throated and short-throated flumes such as the conventional Venturi flume, the Khafagi-Venturi and the Parshall flume.

Where water flows through circular canals with a free water level (no pressure conduits) several types of flumes may be applied such as Venturi's and Palmer Bowlus flumes.

The same structures are used to measure waste water discharge in industrial plants.

### 6.5.3 *Hydrological studies*

Both in hydrological studies and on behalf of the water management in urban and rural areas, many types and different sizes of flow measurement structures are being used.

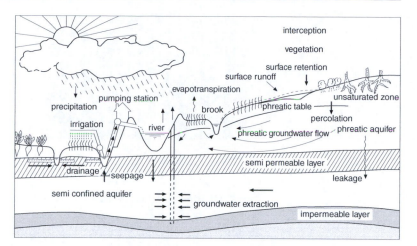

Figure 6.25. The hydrological cycle.

Measurement of surface runoff as a component of the water balance in a hydrological cycle of a catchment area may take place in very small creeks and brooks as well as in large rivers (Figure 6.25).

The variety in types depends mainly on the following conditions:
– the expected discharge range $\gamma = Q_{max}/Q_{min}$
– presence of sediment transport
– available fall
– flow measurement or measurement and regulation (function).

All different types of weirs, flumes and gates are applied in hydrological networks. Sometimes compound weirs are designed. These structures normally consist of a number of overflow structures with different crest-levels and separated by piers.

### 6.5.4 *Laboratory measurements*

The main features of laboratory weirs are:
– the required high accuracy ($X_Q \leq 1\%$)
– sufficient available fall to operate under free flow conditions
– relatively low discharges ($Q_{max} < 100$ to $200$ l/s).

The most appropriate structures are found in the family of sharp-crested weirs:
– horizontal weirs (Rehbock) and rectangular weirs with side contraction
– V-shaped weirs (for $\alpha = 90°$ also called Thomson weir, Figure 6.26)
– circular shaped weirs.

The reason for the high accuracy of sharp-crested weirs is the well de-fined flow pattern and the thorough calibration in several hydraulic laboratories.

Fig. 6.26. A Thomson weir.

## 6.6  DEFINITION OF WEIR FLOW

*Flow measurement structures:* Fixed weirs, adjustable weirs, undershot gates and flumes control both water discharge and water level.

Each discharge measurement structure aims at a local narrowing of the cross section, in which the major part of the total energy-head $H_1$ is converted into kinetic energy used to obtain critical flow, whereas a minor part is lost due to friction at the structure, and by eddies upstream and downstream of the structure.

For all discharge measurement structures a head-discharge relation can be derived. The head is defined as the difference between water level and crest level, where the water level needs to be measured at a sufficient distance upstream of the weir to avoid the influence of the surface drawdown.

Once the relation between the upstream head $h_1$ and the discharge $Q$ has been determined with a certain accuracy, the structure is called a discharge measurement structure or flow measurement structure.

When it is considered necessary also to regulate the flow or the water level, the crest level then must be made movable, thus creating a measurement and regulating structure. Frequently occurring structures are the vertical sliding structures – like the Hobrad weir – and the structures turning around a low-situated or a high-situated horizontal axis. Flow of water over a weir or flume is called *overflow*, whereas flow of water through a submerged opening is called orifice-flow or *underflow*, or gate flow.

In view of the tail water level, distinction should be made between free flow and submerged flow. Discharge under *free flow* conditions supposes a unique relation between the upstream head and the discharge, not depending on the downstream water level. Free flow turns into *submerged flow* as soon as the downstream water level will affect the unique relation between the upstream water level and the discharge.

The submergence ratio is expressed as $S = 100\, H_2/H_1 \approx 100\, h_2/h_1$. The transition between free flow and submerged flow is called the modular limit $S_1$. Free flow occurs for $S < S_1$ and flow becomes submerged for $S > S_1$. The head-discharge equation for horizontal overflow structures – under free flow conditions – reads:

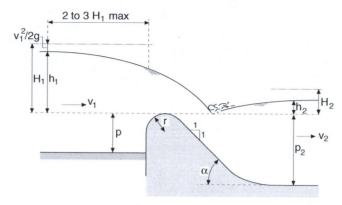

Figure 6.27. Definition sketch of weir flow (after: Bos, 1989).

$$Q = (2/3)^{3/2} \cdot (g)^{1/2} \cdot b \cdot C_{\mathrm{D}} \cdot C_{\mathrm{v}} \cdot h_1^{1.50} \qquad (6.1)$$

where
$Q$  =  discharge (m³/s)
$g$  =  acceleration of gravity $g = 9.81$ m/s²
$b$  =  width of the weir (m)
$C_D$  =  characteristic discharge coefficient (–)
$C_V$  =  coefficient for the approach velocity (–)
$h_1$  =  upstream head over the weir (m)

The general equation reads $Q = C \cdot h_1^{\mathrm{u}}$ where u varies from 0.5 to 2.5 depending on the type and shape of the structure.

The relevant equations for flow measurement structures are given in handbooks, International Standards and other references.

Section 6.7 gives these equations for a restricted number of well-known cross sections.

## 6.7 HEAD-DISCHARGE EQUATIONS

*Broad-crested weirs, the majority of the short-crested weirs, and the flumes*

The symbols used in table 6.4 are explained as follows:
$Q$  =  discharge (m³/s)
$g$  =  acceleration of gravity $g = 9.81$ m/s²
$b$  =  crest width of the weir (m)
$C_D$  =  characteristic discharge coefficient (–)
$C_V$  =  approach velocity coefficient (–)
$h_1$  =  measured head (m)
$\alpha$  =  bottom angle V-shape
$y_c$  =  critical depth in the control section, $y_c = f(H_1/b, m)$
$m$  =  side slope control section (–)

Table 6.4. Head discharge equations for broad-crested weirs.

| Shape in control section | | Head-discharge equation for free flow | Eq. |
|---|---|---|---|
| | rectangular | $Q = \left(\dfrac{2}{3}\right)^{3/2} * g^{1/2} * b * C_D * C_V * h_1^{1.50}$ | (6.2) |
| | V-shaped | $Q = \left(\dfrac{4}{5}\right)^{5/2} * \left(\dfrac{g}{2}\right)^{1/2} * tn\left(\dfrac{\alpha}{2}\right) * C_D * C_V * h_1^{2.50}$ | (6.3) |
| | trapezoidal | $Q = C_D * (by_c + my_c^2) * \{2g(H_1 - y_c)\}^{1/2}$ | (6.4) |
| | V-shaped more than full | $Q = \left(\dfrac{2}{3}\right)^{3/2} * g^{1/2} * B_c * C_D * C_V * (h_1 - 0.5\,H_b)^{1.50}$ | (6.5) |

$H_1$ = upstream energy-head (m)
$H_b$ = height of bottom triangle (m)

The characteristic discharge coefficient $C_D$ is a function of the geometry of the weir-crest. Handbooks and International Standards give additional information about the limits of application and the values of $C_D$.

The approach velocity coefficient is expressed by

$$C_V = \left[\frac{H_1}{h_1}\right]^u \qquad (6.6)$$

where u equals the power of $h_1$ in the head-discharge equation, depending on the shape of the control section.

Values of $C_V$ as a function of the area ratio $C_D A^*/A_1$ are shown in Figure 6.28 for various control sections, where $A^*$ equals the imaginary wetted area at the control section if we assume that the water depth

Table 6.5. u-powers depending on the shape of the control.

| Shape of control section | u-power |
|---|---|
| Rectangular | 1.5 |
| Parabolic (circular) | 2.0 |
| V-shaped | 2.5 |
| V-shaped 'more than full' | 1.5 < u < 2.5 |

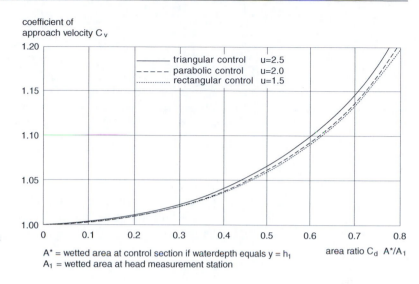

Figure 6.28. Approach
velocity coefficient for various
control sections (after: Bos,
1989).

$y = h_1$; $A_1$ equals the wetted area at the head measurement station and $C_D$ is the discharge coefficient.

*Sharp-crested weirs and some short-crested weirs*
The head-discharge equations for a number of well-known sharp-crested weirs are given in Table 6.6.

*Gates*
Flow through gates is also called underflow and in some cases sluice gate flow (Figure 6.29).

The head-discharge equations are:

*free flow*

$$Q = C_1 \cdot b \cdot a \cdot \sqrt{2gh_1} \qquad\qquad (6.7)$$

Table 6.6. Head-discharge equations for sharp-crested weirs.

| Shape of control section | | Head-discharge equations for free flow | Eq. |
|---|---|---|---|
| | rectangular | $Q = C_e * \dfrac{2}{3}\sqrt{2g} * b * h_1^{1.50}$ | (6.8) |
| | triangular V-shaped | $Q = C_e * \dfrac{8}{15}\sqrt{2g} * tn\left(\dfrac{\alpha}{2}\right) * h_1^{2.50}$ | (6.9) |

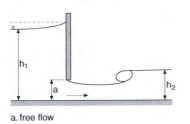 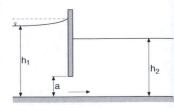

a. free flow                               b. submerged flow

Figure 6.29. Flow under a sluice gate.

where

$a$ = gate opening (m)
$C_1$ = characteristic discharge coefficient for free flow (–)
$C_1$ is a function of the parameter $h_1/a$ and the bottom shape of the gate

*submerged flow*

$$Q = C_2 \cdot b \cdot a \cdot \sqrt{2g(h_1 - h_2)} \qquad\qquad (6.10)$$

where

$h_2$ = downstream head
$C_2$ = coefficient for submerged flow
$C_2$ depends on the parameter $h_2/h_1$ and the bottom shape of the gate

## 6.8  SELECTION OF THE MOST SUITABLE FLOW MEASUREMENT STRUCTURE

The selection of a structure and the design of its dimensions determine to a high degree the quality of the discharge measurements. The designer will make his choice on the basis of the characteristics of the structures, the field or boundary conditions, and the human requirements (demands) imposed by the water management.

### 6.8.1  *Characteristics*

The characteristics of the numerous different structures are expressed in a number of properties:
– *loss of head required by the measuring device*
   Structures with a high discharge capacity are characterized by a high discharge coefficient. For example, the short-crested weir with cylindrical crest $C_D = 1.48$ needs considerably less head than the broadcrested rectangular profile weir $C_D = 0.85$. On the other hand, the necessary loss of head is decreasing with a higher modular limit, consecutively for sharp-crested weirs (a low value) and flumes (a high value).

- *measuring range*
  The shape and the width of the crest both determine the minimum discharge, assuming a minimum head $h_1 = 0.05$ m. The measuring range is defined as $\gamma = Q_{max}/Q_{min}$.
  Structures with a triangular cross-section allow larger ranges than structures with a rectangular cross-section.
- *ability to transport solid materials*
  The passage of sediment across the bottom of the approach channel will be facilitated by a low crest height and a streamlined inflow (flumes). Gates and overflow structures with low sills (or even without a sill) are the most favourable structures with respect to sediment transport capability. Transport of floating materials needs a streamlined structure, including the crest shape: sharp-crested devices will for this reason not be selected as discharge measuring structures in water in which there is floating debris.
- *sensitivity*
  The overall error in flow measurements with structures depends strongly on the sensitivity $S$, which is expressed as the variation of the discharge caused by a relative change in the upstream water level.

$$S = u \cdot \Delta h/h_1 \tag{6.11}$$

where
$\Delta h/h_1$ = is the relative change in upstream water level
$u$ = the power in the equation $Q = C \cdot h_1^{u}$

Gates are by far least sensible to water level changes, whereas overflow structures are three to five times more sensitive.
- *accuracy*
  The accuracy of the structure depends on the number and the reliability of the calibrations and whether the measurements can be reproduced within a limited percentage. Sharp-crested weirs are famous for their high accuracy.

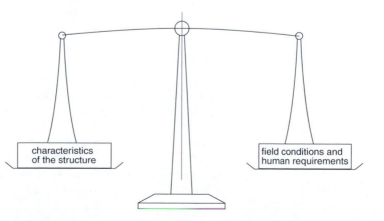

Figure 6.30. Selection of the type of flow measurement structure.

— *possibilities to regulate discharges or water levels*
   Some structures have been developed both as a measurement and regulating structure, either by movability in vertical slots or by turning
   around an axis. Other structures are exclusively designed with a fixed
   crest (flumes).

### 6.8.2  *Field conditions*

The selection of the location of the measurement station and the type of
structure will be influenced by the field or boundary conditions. The most
relevant information will be given by:
— *the available head*
   The installation of a structure causes a loss of head which must be
   created by raising the upstream water level or by lowering the downstream water level. Both actions have their limitations.
— *the ranges of discharges and water levels*
   In the case of designing a structure in a natural stream, the designer
   should be informed about the range of discharges and water levels and
   about the frequence of occurrence.
— *transport of solid material*
   Natural streams as well as artificial and recently designed canals may
   transport sediment or floating debris. Both are seasonally-dependent
   and will mostly occur during a limited period, but in a high degree.

### 6.8.3  *Human requirements*

From the water management point of view the following demands have
to be considered:
— *requirements covering the function of the structure*
   necessity of regulating flow or water levels. In many cases we want
   to regulate the water supply and to control the upstream or the downstream water level.
— *required minimum water level*
   To maintain a minimum water level, for example during drought periods, the crest level of overflow structures shall either lie at a certain
   height (apex height) above the canal bottom or be adjustable (movable
   weir), which also can be done using gates.
— *flexibility*
   Figure 6.31 shows how the incoming flow $Q$ from a supply canal is
   bifurcated over the offtake canal $Q_o$ and the continuing supply canal
   $Q_s$. The discharges $Q_o$ and $Q_s$ are measured with hydraulic structures. If for any reason the upstream water level would change (as
   a consequence of a change in the incoming flow $Q$), the discharges
   $Q_o$ and $Q_s$ will change also. Will they change by the same percentage (proportionally)? To answer this question the term flexibility $F$
   is used.

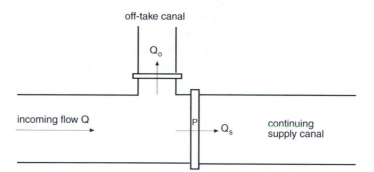

off-take canal

$Q_o$

incoming flow Q

P

$Q_s$

continuing
supply canal

Figure 6.31. Flexibility at a
bifurcation.

$$F = \frac{\text{rate of change of } Q_o}{\text{rate of change of } Q_s} = \frac{\dfrac{dQ_o}{Q_o}}{\dfrac{dQ_s}{Q_s}} \qquad (6.12)$$

With the general rating for a structure $Q = C \cdot h_1{}^u$, the flexibility can be rewritten as:

$$F = u_o \cdot h_s / u_s \cdot h_o \qquad (6.13)$$

where:
$u_o$ = u-power of the structure in the off-take canal
$u_s$ = u-power of the structure in the continuing supply canal
$h_o$ = upstream head over the structure in the off-take canal
$h_s$ = upstream head over the structure in the continuing supply canal

There are three possibilities with respect to the flexibility of the bifurcated offtake (assume free flow at both structures):
1. $F = 1$
   This holds when both structures are of the same type $u_o = u_s$ and when they have the same crest level or sill level. Distribution is proportional and not depending on upstream changes.
2. $F < 1$
   This happens for instance when the offtake structure is an under-shot gate, $u_o = 0.5$ and the continuing canal has a horizontal over-flow structure, $u_s = 1.5$.
   $F = 0.5 \, h_s / 1.5 \, h_o$
   provided $h_s < 3h_o$, then $F < 1$.
   With this design the variation in the offtake is less than in the continuing canal, which is an advantage for tail-end users of irrigation water in the offtake canal.
3. $F > 1$
   Now assume the offtake structure is a horizontal overflow structure, $u_o = 1.5$ and the continuing canal has an undershot gate $u_s = 0.5$.

$F = 1.5 \, h_s / 0.5 \, h_o$

provided $h_o < 3h_s$, then $F > 1$.

This may occur when the offtake structure acts as a spillway, to prevent the supply canal from overload.

For any bifurcation, the flexibility of the offtake can be calculated from the u-values (type of structure) and the *h*-values (water level related to sill level).

– *accuracy*
Field structures will measure discharges generally with an error $X_Q \approx$ 5%. The accuracy in the evaluation of discharges is dealt with in Section 6.11.
– *non-technical demands such as:*
  – the availability of construction materials
  – familiarity with a certain type of structure
  – importance of standardization.
    It is generally recommended that the number of different types of structures in an area be restricted. When new types are introduced, technicians and farmers need to be familiarized. Standardization leads also to a reduction in costs.
  – at many places structures are exposed to alterations by unauthorized people. Such alterations can be avoided by building the structures sturdily and by locking movable parts.

The most suitable structure can only be selected after performing an in-depth study into the integration of all field-boundary conditions and human requirements.

## 6.9  DESIGN OF A FLOW MEASUREMENT STRUCTURE

### 6.9.1  *Introduction*

In Section 6.9 the design of a flow measurement structure is discussed:
– selection of the most appropriate type, based on field conditions and human requirements (6.9.2)
– description of the selected type (6.9.3)
– hydraulic design of the weir (6.9.4)
– set up of the rating curve (6.9.5)

### 6.9.2  *Selection of the most appropriate type*

*Field conditions*
– available information about the channel
  – bottom width $b = 10$ m
  – side slope 1 vertical: 2 horizontal (m = 2)
  – field level 3.20 m above channel bottom

- overall roughness bottom and side slopes $n = 0.035$ (maximum value)
- longitudinal slope of channel $S = 2.10^{-4}$
- water level not affected by downstream restrictions (normal depth)
- additional boundary conditions:
  - expected range of discharges $1$ m³/s $< Q < 15$ m³/s
  - available fall (loss of head) $\Delta h = 0.60$ m for $Q_{max} = 15$ m³/s
  - transport of solid material: no bed load transport, floating debris is expected.

*Human requirements*
- no need to regulate flow or water levels → fixed crest
- desired measuring range $1$ m³/s $< Q < 15$ m³/s
- desired accuracy: $X_Q \leq 7{,}5\%$ for the complete range
- non technical demands: a standardized structure

*Selection of the type*
- discharge range $\gamma = Q_{max} / Q_{min} = 15$ → no need for a triangular or a trapezoidal shaped weir. A rectangular (horizontal) weir gives a good solution
- available fall $\Delta h = 0.60$ m for $Q = 15$ m³/s → a weir or flume is needed with a high modular limit to prevent submerged flow: a broad-crested weir or a long-throated flume
- no bed load transport → no need for a flume: a broad-crested weir
- floating debris is expected → sharp corners to be avoided: a round-nose broad-crested weir between vertical side walls with rounded abutments.

*Conclusion*:    the best type is a round-nose broad-crested weir, which has been standardized in ISO 4374
This structure is described in Section 6.9.3

### 6.9.3 *Description of the round-nose horizontal broad-crested weir*

- discharge equation $Q = \left(\dfrac{2}{3}\right)^{3/2} * g^{1/2} * b * C_D * C_V * h_1^{1.50}$

  with $C_D = f(H_1/L)$

- *sensitivity factor* $u = 1.5$

- *movability of the crest* $\begin{cases} \text{weirs with fixed crest (Figure 6.32)} \\ \text{adjustable weirs (Figures 6.33 and 6.34)} \end{cases}$

- *vertical side walls:* indispensable

- *head range* $\begin{cases} 0.10 < H_1/L < 0.40 & \text{broad-crested flow} \\ 0.40 < H_1/L < 1.20 & \text{short-crested flow} \end{cases}$

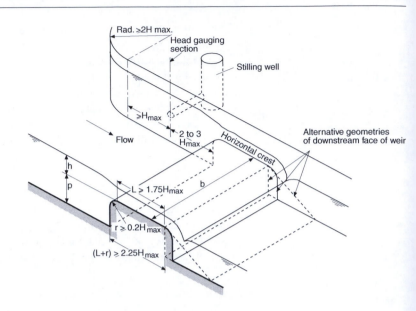

Figure 6.32. Definition sketch
round-nose horizontal broad-
crested weir (after: ISO).

– *range of discharge:*

$$\gamma = Q_{max}/Q_{min} \qquad \begin{array}{l} \gamma = 8 \text{ for broad-crested flow} \\ \gamma = 45 \text{ including short-crested flow} \end{array}$$

– *accuracy of $C_D$-values:* $X_c = 2.0 + 0.15\, L/H_1\ (\%)$

– *required fall:*

the modular limit $\begin{cases} S_1 = 67\% \text{ for } H_1/L \leq 0.40 \\ S_1 < 67\% \text{ for } H_1/L > 0.40 \text{ (Table 6.7)} \end{cases}$

– *sediment passing capacity:*
poor for weirs with fixed crest
good for adjustable weirs with bottom slide

– *floating debris capacity:* good

– *state of standardization:* International Standard ISO 4374

– *references:*    Round-nose horizontal broad-crested weirs with fixed
crest and with movable crest (Boiten, 1987)

– *application:*

$\begin{cases} \text{weirs with fixed crest : in natural streams} \\ \text{adjustable weirs : in irrigation canals as offtake structure} \end{cases}$

– *remarks:* additional information is given in ISO 4374

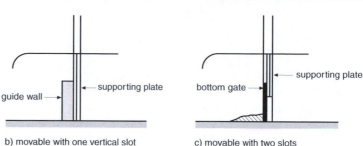

a) with fixed crest           b) movable with one vertical slot       c) movable with two slots

Figure 6.33. Three methods of using the weir.

Traditionally the round nose horizontal broad-crested weirs have been designed and constructed as concrete weirs with a fixed crest.

Nowadays, many engineers select the weir as a measurement and regulating structure, made of steel or aluminium and movable in vertical slots. Such a structure is very versatile for use in relatively flat areas where the water demand and the water level may vary during the season.

Figure 6.33 shows three methods of application.

In situations where accumulation of sediment (sand and silt) may be expected, method c) will be applied. This movable weir is also called: the Hobrad-weir (*ho*rizontal *br*oad-crested *ad*justable). Figure 6.34 shows the Hobrad weir.

The International Standard ISO 4374 gives the discharge coefficient $C_D$ as a function of the ratios $h_1/L$ and $b/L$ and depending on the roughness of the crest, for $h_1/L \leq 0.57$ (broad-crested flow).

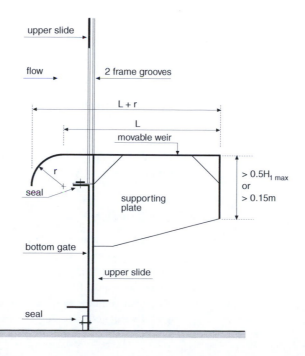

Figure 6.34. The Hobrad weir (after: Boiten, 1987).

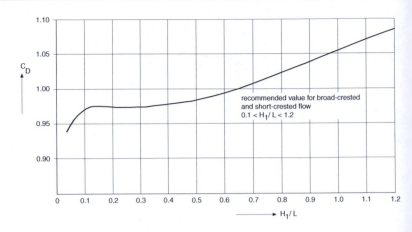

Figure 6.35. Discharge coefficient $C_D$ as a function of $H_1/L$ (after: Boiten, 1987).

In the referred study (Boiten, 1987) the head range has been extended to $h_1/L = 1.20$ so including short-crested flow.

For weirs made of stainless steel, aluminium or painted metal the roughness of the crest will not affect the discharge coefficient.
    Figure 6.35 gives $C_D$ as a function of $H_1/L$.

Limits of application:

$H_{min}^{0.5} \cdot L \geq 0.12$ (to prevent scaling effects)
$H_{max}/p \leq 1.5$
$p \geq 0.15$ m
$b \geq 0.30$ m
$b \geq H_{max}$
$b \geq L/5$

As a consequence of the curved streamlines in the range of short-crested flow the modular limit will decrease for an increasing $H_1/L$ ratio.

For the short-crested flow range, $H_1/L > 0.57$, it is recommended to keep the submergence ratio $S \leq 0$.

### 6.9.4 *Hydraulic design*

Successively the following calculations are carried out:
– the relation between normal water depth $y$ and discharge $Q$ in the channel before constructing the weir
– the main dimensions of the weir:
    the width $b$ between the vertical side walls
    apex height $p$, related to the upstream channel bottom
    crest length $L$ and the round-nose radius $r$

Table 6.7. Modular limit as a function of $H_1/L$ for weirs with a vertical backface.

| $H_1/L$ | Modular limit $S_1$ |
|---|---|
| ≤  0.33 | 67% |
| 0.40 | 60% |
| 0.50 | 50% |
| ≥  0.57 | <  40% |

From Table 6.8 can be seen:

$$Q_{min} = 1 \text{ m}^3/\text{s} \rightarrow y = 0.43 \text{ m}$$

$$Q_{max} = 15 \text{ m}^3/\text{s} \rightarrow y = 2.02 \text{ m}$$

*The main dimensions of the weir*
– *The width b is governed by the following conditions:*
a) $h_{min} \geq 0.15$ m (accuracy for $Q_{min} = 1$ m³/s)
$\qquad Q = 1.705 \cdot b \cdot C_D \cdot C_v \cdot h_1^{1.50}$
$\qquad$ take for the design purposes $C_D \cdot C_v = 1$ and $b \leq 10.10$ m
b) for $Q_{max} = 15$ m³/s: $\qquad$ available fall $\Delta h = 0.60$ m
$\qquad\qquad\qquad\qquad\qquad\qquad$ $S_1 \leq 40\%$ to avoid submerged flow
From the table can be seen: $b \geq 8.80$ m $\qquad (\Delta h \leq 0.60$ m)

Conclusion: $\qquad$ $8.80 < b < 10.10$ m
$\qquad\qquad\qquad$ design width $b = 9.00$ m

– *The apex height p of the weir*
$\qquad$ For $Q_{max} = 15$ m³/s and $b = 9.00$ m, the approximate values are found:
$\qquad$ $h_1 = 0.98$ m and $h_2 = 0.39$ m (Table 6.9)
$\qquad$ From Table 6.8 we found for $Q_{max} = 15$ m³/s a depth $y = 2.02$ m.
$\qquad$ The apex height is (see Figure 6.36) $p = 2.02 - 0.39 = 1.63$ m.

– *The crest length L and round-nose radius r*
$\qquad$ For $Q_{max} = 15$ m³/s and $b = 9.00$ m we found $h_1$ max $= 0.98$ m
$\qquad$ From Table 6.7: $S_1 = 40\%$ holds $H_1/L = 0.57$
$\qquad$ The crest length $L \geq h_1$ max/0.57 = 0.98/0.57 $\qquad L \geq 1.72$ m
$\qquad$ Design: $L = 1.80$ m and $r = 0.20$ m

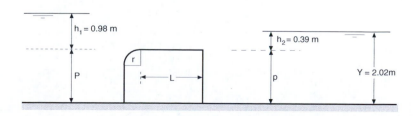

Figure 6.36. Determination apex height $P$.

Table 6.8. Calculation relation between normal depth y and discharge Q.

| Normal water depth y (m) | Perimeter p (m) | Area A (m²) | Hydr. radius R=A/P (m) | $\bar{v} = \frac{1}{n} \cdot R^{2/3} \cdot S^{1/2}$ (m/s) | Discharge Q = v̄ · A (m³/s) |
|---|---|---|---|---|---|
| 0.30 | 11.34 | 3.18 | 0.280 | 0.173 | 0.55 |
| 0.50 | 12.24 | 5.50 | 0.449 | 0.237 | 1.30 |
| 0.75 | 13.35 | 8.62 | 0.646 | 0.302 | 2.60 |
| 1.00 | 14.47 | 12.00 | 0.829 | 0.357 | 4.28 |
| 1.25 | 15.59 | 15.62 | 1.002 | 0.405 | 6.32 |
| 1.50 | 16.71 | 19.50 | 1.167 | 0.448 | 8.73 |
| 1.75 | 17.83 | 23.62 | 1.325 | 0.487 | 11.51 |
| 2.00 | 18.94 | 28.00 | 1.478 | 0.524 | 14.68 |
| 2.25 | 20.06 | 32.62 | 1.626 | 0.559 | 18.23 |

Table 6.9. Calculation minimum width of the weir.

| Width b (m) | $h_1$ max = (15/1.705 b)$^{2/3}$ (m) | $h_2$ max = 0.4 $h_1$ max (m) | $\Delta h = h_1 - h_2$ (m) |
|---|---|---|---|
| 7 | 1.16 | 0.46 | 0.70 |
| 8 | 1.07 | 0.43 | 0.64 |
| 9 | 0.98 | 0.39 | 0.59 |
| 10 | 0.92 | 0.37 | 0.55 |

*The main dimension of the weir are as follows:*
- width $b = 9.00$ m
- apex height $p = 1.63$ m
- crest length $L = 1.80$ m and $r = 0.20$ m

All the limits of application are fulfilled.

### 6.9.5 *Set up of the rating curve*

$$Q = \left(\frac{2}{3}\right)^{3/2} \cdot (g)^{1/2} \cdot b \cdot C_D \cdot C_v \cdot h_1^{1.50}$$

with $g = 9.81$ m/s² and $b = 9.00$ m the relation is:

Table 6.10. Calculation of the rating curve $Q = f(h_1)$.

| $h_1$ (m) | $H_1/L \approx h_1/L$ (–) | $C_D$ (–) | $Q(C_V = 1)$ (m³/s) | A (m²) | $C_V$ (–) | Q (m³/s) |
|---|---|---|---|---|---|---|
| 0.15 | 0.083 | 0.966 | 0.861 | 24.14 | 1.001 | 0.86 |
| 0.25 | 0.139 | 0.974 | 1.868 | 25.87 | 1.002 | 1.87 |
| 0.40 | 0.222 | 0.974 | 3.780 | 28.54 | 1.003 | 3.79 |
| 0.60 | 0.333 | 0.974 | 6.944 | 32.25 | 1.006 | 6.99 |
| 0.80 | 0.444 | 0.979 | 10.746 | 36.11 | 1.008 | 10.84 |
| 1.00 | 0.556 | 0.989 | 15.171 | 40.13 | 1.011 | 15.34 |

Table 6.11. Calculation of the required headloss $y_1 - y_2$ max.

| $h_1$ (m) | $Q$ (m³/s) | $H_1/L \approx h_1/L$ (–) | $S_\ell$ (–) | $h_2$ max (m) | $y_2$ max (m) | $y_1$ (m) | $y_1 - y_2$ max (m) |
|---|---|---|---|---|---|---|---|
| 0.165 | 1.00 | 0.09 | 0.67 | 0.11 | 1.74 | 1.80 | 0.06 |
| 0.250 | 1.87 | 0.14 | 0.67 | 0.17 | 1.80 | 1.88 | 0.08 |
| 0.400 | 3.79 | 0.22 | 0.67 | 0.27 | 1.90 | 2.03 | 0.13 |
| 0.600 | 6.99 | 0.33 | 0.67 | 0.40 | 2.03 | 2.23 | 0.20 |
| 0.800 | 10.84 | 0.44 | 0.57 | 0.46 | 2.09 | 2.43 | 0.34 |
| 0.991 | 15.00 | 0.56 | 0.41 | 0.41 | 2.04 | 2.62 | 0.58 |

$$Q = 15.34 \cdot C_D \cdot C_V \cdot h_1^{1.50}$$
$$A = (h_1 + 1.63) \cdot \{10.00 + 2(h_1 + 1.63)\}$$

The computation of $C_V$ and $Q$ is a short iteration process (for low $C_V$-values, as in this example, the process is only one step).

The relation between the upstream head $h_1$ and the discharge $Q$ for this weir can easily be presented by:

$$Q = 15.2 \cdot h_1^{1.51}$$

From this equation can be seen:

$$Q = 1 \text{ m}^3/\text{s} \rightarrow h_1 = 0.165 \text{ m}$$
$$Q = 15 \text{ m}^3/\text{s} \rightarrow h_1 = 0.991 \text{ m}$$

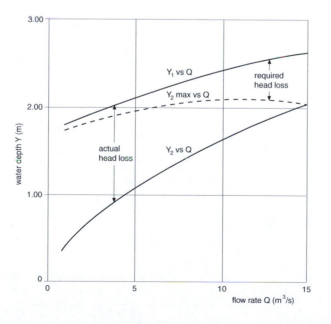

Figure 6.37. Relationship $y$ vs $Q$.

The required head loss – to prevent submerged flow – has been calculated in Table 6.11.

$S_1$ values have been interpolated from Table 6.7

$h_2 \max = h_1 \cdot S_1$

$y_2 \max = h_2 \max + p$

$y_1 = h_1 + p$

$y_2$ values have been interpolated from Table 6.8

$y_1 - y_2 \max$ is the required head loss

Figure 6.37 shows the relationship $y$ vs $Q$, comparing the required head loss and the actual head loss (prevention of submerged flow).

## 6.10  CALIBRATION OF FLOW MEASUREMENT STRUCTURES USING SCALE MODELS

Flow measurement structures can be calibrated in a hydraulic or scale model. A scale model is a reduction of a field structure, built with the aim to investigate certain phenomena as flow pattern, local scour, head discharge relations etc.

*Advantages:* independent from field conditions (weather and unsteady flow), accuracy of model measurements, costs, especially for design purposes.

1. *Model scales n*

$$\text{length scale } n_\ell = \frac{\text{prototype length } (1_p)}{\text{model length } (1_m)} \tag{6.14}$$

generally the length reduction equals the depth reduction → non-distorted scale models (sometimes it is necessary to give more fall to a river model in order to get the beginning of motion of bed material; then a distorted scale model is made).

2. *Rules for calculation of scales*
   if $x^a = y^b \cdot z^c$ ($x$, $y$ and $z$ have dimensions; $a$, $b$ and $c$ are components) than
   $(n_x)^a = (n_y)^b \cdot (n_z)^c$

   for example: $h_v = v^2/2g$    $h_v$ = velocity head
   $v$ = velocity
   $g$ = gravitational acceleration

   $n_h = (n_v)^2/n_{2g}$
   $n_h = (n_v)^2/1$
   $n_h = n_v^2 \rightarrow$ velocity scale $n_v = (n_h)^{1/2} = (n_\ell)^{1/2}$ \hfill (6.15)

discharge    $Q = v \cdot A$    $A = $ length $\times$ depth

$$n_Q = n_v \cdot n_A$$
$$n_Q = (n_\ell)^{1/2} \cdot (n_\ell)^2$$

discharge scale $n_Q = (n_\ell)^{5/2}$ \hfill (6.16)

3. *Scaling effects*

Froude law      $Fr = \dfrac{v}{\sqrt{gA/b}}$ \hfill (6.17)

where
$Fr$ = Froude number (–)
$v$  = velocity (m/s)
$g$  = gravitational acceleration (m/s)
$A$  = cross-sectional area (–)
$b$  = width on the water surface (m)

Under normal conditions yields $n_{Fr} = 1$, assuming the motion of flow is determined exclusively by gravitational acceleration, and neglecting viscosity and surface tension.

*Example:* Calibration of a weir in the range 0.05 m $< h_1 <$ 1.50 m where $h_1$ is the expected head in prototype.
Calibration of the weir with a scale model $n_1 = 5$ gives
0.01 m $< h_{model} <$ 0.30 m.
For $h_{model} = 0.01$ m viscosity and surface tension will affect the head discharge relation by 2%–10% approximately (percentage comes from laboratory experience). Especially weirs with a curved crest (circular, parabolic) are rather sensitive for scaling effects.
The minimum scale factor to prevent scaling effects depends on the shape of the weir. For round-nose broad-crested weirs, scaling effects are substantially prevented for length Reynolds numbers $Re_L > 3.10^5$.

$$Re = \frac{v \cdot L}{v}$$ \hfill (6.18)

where
$Re$ = Reynolds number (–)
$v$  = velocity above the crest (m/s)
$L$  = crest length (m)
$v$  = kinematic viscosity (m²/s)
To prevent scaling effects in the above example two models are needed:
$n_\ell = 1$   0.05 m $<$ h$_1$ $<$ 0.30 m        model 0.05 $< h_{mod} <$ 0.30 m
$n_\ell = 5$   0.25 m $< h_1 <$ 1.50 m        model 0.05 $<$ h$_{mod}$ $<$ 0.30 m

4. *Conditions, governing the selection of a model scale*
   – available area in the laboratory
   – capacity of discharge in the laboratory
   – desired accuracy of measurement results
   – scaling effects
   – cost of construction of the model

## 6.11  FISHWAYS

### 6.11.1  *Classification of fishways*

Fish migration in natural streams may be restricted if a structure fails to make proper provision for their passage. A flow measurement structure may be an obstruction for fish migration.

The principal factors affecting the passage of fish through a structure are the flow velocities in and downstream of the critical section, the available width and depth of flow at the structure and the drop in the water level over the structure.

The suitability of flow-measurement structures for fish passage depends on:
– the migratory species: swimming capabilities and leap height
– the type of structure, flow velocities and drop in water level

In recent years, local and national water authorities are attempting to reestablish fish migration in their rivers.

*Definition, purpose and classification of fishways*
Fishways can be *defined* as hydraulic structures, which facilitate upstream and downstream migration of certain fish species in rivers where natural or man-made obstructions – dams, hydropower stations, pumping stations, water mills and weirs – prevent a free migration.

The main *purpose* of fishways is to provide acceptable flow conditions for migrating fish. Acceptable flow conditions are defined in terms of the flow pattern, flow velocities, a desired drop in water level, resting places, etc.

Most fishways are constructed in a bypass channel parallel to the main water course, in which a hydraulic structure has been constructed.

Fishways are *classified* as follows:
– *Pool and weir fishways*, used since the 19th century
  These fishways consist of a sloping or stepped channel (rectangular or trapezoidal cross section) partitioned into a number of pools by weirs. Flow over the weirs is plunging (modular flow) or streaming (non-modular flow). The energy is dissipated in the pools, provided the pools have sufficient volume.
  Pool and weir fishways can be subdivided into many subtypes, mainly depending on the weir geometry, such as: central rectangular notches, lateral notches, central V-shaped weirs.

Table 6.12. Suitability of flow measurement structures for fish migration.

| Type of structure | Suitability |
| --- | --- |
| Sharp-crested weirs and end depth methods | Very poor (aerated nappe) |
| Broad-crested weirs, short-crested weirs, gates, Parshall and Saniiri flumes | Poor, depending on submergence ratio $S$ |
| Long throated and most short throated flumes | Rather good, especially for throats without bottom hump |

– *Denil fishways,* designed since 1908

Denil fishways consist of a sloping rectangular channel with closely spaced baffles on the sides and the bottom. Due to high-energy dissipation, flow is highly turbulent. The baffles are placed vertically or under certain inclination.

– *Vertical slot fishways*

These fishways consist of a sloping or stepped rectangular channel, partitioned into a number of pools, separated by vertical open slots. Vertical slot fishways can be subdivided into many subtypes, such as: single jet slots one-sided, single jet slots alternatively sided, double jet slots, and various other types.

### 6.11.2 *Rating Curves of fishways*

Satisfactory hydrometric calibrations are available for the following
　　fishways
– super-active Larinier fishpass
– pool-type fishway with V-shaped overfalls (Boiten, 1990).
– the Dutch pool and origin fishway (Boiten and Dommerholt, 2006)
ISO/TC 113 is preparing an international standard ISO 26906. Fishpasses
at flow measurement structures.

Figure 6.38. A pooltype
fishway in The Netherlands.

The main purpose of all these fishways is fish passage. Nevertheless, for all of them head-discharge relations can be determined.

The availability of a rating curve for the fishway is important in those situations where a fishway has been constructed in a bypass parallel to a flow-measurement structure in the main course, because most of the discharge during dry periods may pass through the fishway.

In some cases a rating curve of a fishway can be derived from its design rules. The best accuracy will be achieved by a calibration of the fishway using a scale model.

## 6.12  ACCURACY OF DISCHARGE MEASUREMENTS

Errors occurring in indirect discharge measurements have different natures. The most significant contributions are:

1. The sensitivity of the structure: change of discharge per unit change of head. The sensitivity is proportional to the u-power in $Q = C \cdot h_1^u$, consequently on the shape of the cross section.
2. The way in which the head $h_1$ is measured and the method of registration of both the water level and the crest level.
3. Errors in the calibration affecting the reliability of the characteristic discharge coefficient $C_D$.
4. Errors in the dimensions of the structure ($B$ and $\alpha$).

The magnitude of the random error in the determination of the discharge is derived from the head-discharge relation $Q = C \cdot B \cdot h_1^u$

$$X_Q = \sqrt{(X_C^2 + X_B^2 + (uX_{h_1})^2)} \qquad (6.19)$$

| | |
|---|---|
| gates | u = 0.5 |
| sutro weir | u = 1.0 |
| rectangular cross section | u = 1.5 |
| parabolic cross section | u = 2.0 |
| V-shaped cross-section | u = 2.5 |

$X_Q$  error in discharge (%)

$X_C$  error in $C_D$, generally $2\% < X_C < 5\%$

$X_B$  error in width, can be corrected (%)

$X_h$  error in head measurement $X_h = 100(\delta_h/h_1)$  (%)   (6.20)

$\delta_h$  absolute error in head measurement, normally $0.002 \text{ m} < \delta_h < 0.005 \text{ m}$.

A more detailed description of the uncertainties in discharge estimation, including examples, is given in the ISO standards. (Section 6.13).

## 6.13  STANDARDIZATION OF FLOW MEASUREMENT STRUCTURES

The International Organization for Standardization (ISO) issued by its Technical Committee TC 113 a number of Standards on flow measurement structures. Table 6.13 gives a summary of the ISO-standards, which have been drafted by experts in many countries all over the world.

Table 6.13. Flow measurement structures, standardized by ISO/TC 113.

| Type of structure | Structure name | ISO Standard |
|---|---|---|
| Broad-crested weirs | Round-nose horizontal broad-crested weirs | ISO 4374-1990 |
| | Rectangular broad-crested weirs | ISO 3846-1989 |
| | Trapezoidal profile weirs | ISO 4362-1999 |
| | V-shaped broad-crested weirs | ISO 8333-1985 |
| Sharp-crested weirs | Thin plate weirs | ISO 1438-1980 |
| Short-crested weirs | Triangular profile weirs | ISO 4360-1984 |
| | Flat V-weirs | ISO 4377-2002 |
| | Streamlined triangular profile weirs | ISO 9827-1994 |
| End depth methods | Rectangular channels with a free overfall | ISO 3847-1977 |
| | Non-rectangular channels with a free overfall | ISO 4371-1984 |
| Flumes | Rectangular, trapezoidal and U-shaped flumes | ISO 4359-1983 |
| | Parshall and SANIIRI flumes | ISO 9826-1992 |
| Gates | Vertical underflow gates | ISO 13550-2002 |
| All types | Guide-lines for the selection of flow measurement structures | ISO 8368-1999 |
| | Compound gauging structures | ISO 14139-2000 |
| | Fishpasses at flow measurement structures | ISO 26906 |

CHAPTER 7

# Hydrological networks

## 7.1 INTRODUCTION

Measurement of hydrological parameters dates back many centuries. In early times daily readings from single water level gauges were taken for the benefit of agriculture or navigation. Later on the first hydrological networks were set up, in which various parameters were measured (water levels, discharges and rainfall) which were related to one another. Nowadays hydrological networks are installed in large and small watersheds, whereas the field data are collected automatically and transmitted by telemetry systems. In addition to the usual water quantity data, in many cases water quality parameters (oxygen, pH, nutrients and others) are measured as well.

*The evolution in measuring hydrological parameters*
South of Cairo we still find the Roda Gauge in the river Nile. Observations from this water level gauge are known since 641 A.D. Every year at the end of a flood-period the egyptian finance minister related the tax-rates for farmers to the maximum water level.

Since 1770 the first water level measurements were carried out along the river Rhine, at various locations in order to get a better insight into the behaviour of the river.

In the same period hydrological measurements were performed in the watershed of the river Seine in France. River flow and rainfall were measured, and related to one another.

So, there was a growing interest of hydrologists in the coherence of measurements of various parameters and on various locations. The first hydrological networks were set up in large catchment areas. Nowadays water managers work with integrated hydrological networks in both large and small catchments. The raw field data from measurements on surface water levels, groundwater depths, rainfall and river discharges are elaborated, validated and analysed by modern database management and processing systems. The final hydrological information is used both for water management purposes and hydrological research among which is the set up of water balances.

Table 7.1. Water management by public authorities in the Netherlands.

| Object of management Management level | Surface water National waters | Regional and local waters | Groundwater |
|---|---|---|---|
| Strategic level | Central government | Provinces | Provinces |
| Operational level | Central government | Waterboards | Provinces and waterboards |

*Scheme of water resources management in the Netherlands*
In the Netherlands three levels of government have responsibilities in water management as indicated in Table 7.1: the central government, the 12 provinces and the 26 regional waterboards.

At these three levels the authorities make their own water resources management plans. At the strategic level it is a masterplan and at the operational level the plans are more detailed.

In the field of hydrological networks, the majority of the waterboards is very active. The governing boards of the 26 waterboards are elected by landowners and other groups. Waterboards are responsible for 100.000 to 150.000 hectares of land area on the average. These boards have the first responsibility for the water management in their region. Therefore they make their own water management plan, including the *hydrological plan*, which is represented schematically in Table 7.2.

The *hydrological plan* includes:
– the hydrological network: a comprehensive network of gauging stations for the measurement of water levels, groundwater depths, discharges and rainfall, in some cases completed with measurements of water quality, sediment transport and riverbed sediment quality.
– all technical and administrative facilities and a well qualified staff to collect and to elaborate the measured data and to use all information in an appropriate and functional way.
    In the Netherlands water quality management at the regional level is also the responsibility of the waterboards.
A hydrological network, operating satisfactorily, is an indispensable tool for a reliable watershed management.

Table 7.2. The water management plan of a waterboard.

|  | Province (supervision) ⇓ |  |
|---|---|---|
| Catchment characteristics and various interests ⇒ | Water management plan waterboards ⇑ | ⇒ Policy and management |
|  | *Hydrological plan* |  |

*Basic networks*

A hydrological network is a set of stations at which hydrological observations are made over time. The primary objective is to make an inventory of the water resources for general planning purposes.

Three categories of stations can be identified:
- principal stations, for a long-term use
- secondary stations to be operated for a couple of years
- stations for specific purposes: flood forecasting, irrigation, water supply and urban development

In this chapter the following items are discussed
- the dynamics of a water system (Section 7.2)
- purpose and first setting-up of a hydrological network (Section 7.3)
- optimization of monitoring networks (Section 7.4)

## 7.2 THE DYNAMICS OF A WATER SYSTEM

Hydrological systems are three-dimensional and time-dependent. They are characterized by spatial and temporal variability. In such a system distinction is made between time-independent characteristics of the area, season-dependent characteristics of the area and variable hydrological and meteorological parameters. The behaviour of the hydrological variables is strongly related to the characteristics of the catchment-area.

*Time-independent characteristics* of the area are:
- geology: what sedimentary deposits are present and at what location,
- geohydrology: how is the subsoil structure in relation to aquifers,
- topography: how is the surface relief and what is the height of the area,
- soil science: soil type, soil layers, hydraulic properties, infiltration capacity and storage capacity,
- drainage: how is the drainage pattern of the area.

*Season-dependent characteristics* of the area are:
- variation in the supply of energy (sunshine),
- variation in land-use and vegetation.

*Time-dependant hydrological variables* playing a prominent role in the system, are:
- water levels in the open channels,
- discharges,
- groundwater depths,
- precipitation such as rain or snow,
- evaporation of open water and evapotranspiration of overgrown areas,
- moisture content of the soils,
- a number of water quality parameters: acidity pH and oxygen content.

All these hydrological variables can be measured in a hydrological network: a number of these parameters is measured in a number of locations and with a certain frequency. Only the evapotranspiration needs to be calculated.

The density of a network depends on the spatial variation whereas the frequency is determined by the temporal variation of the parameter.

The following three examples demonstrate the spatial and temporal variations in hydrological systems.

1. Precipitation and evaporation vary in time and place: variation in precipitation is strong, variations in evaporation are relatively weak. As a consequence the frequency of readings will be different for both parameters.
2. Groundwater table variations depend on precipitation, evaporation, soil type, and drainage. In addition they are affected by the tidal motion along the coast and along tidal rivers, as well as by barometric variations.
3. The variation in river discharges from different catchment areas is governed by the magnitude of the base flow component and by the storage capacity in the area (soil type, vegetation, slope, urbanization), and by the rainfall.

The general rule for getting a reliable insight into the behaviour of a water system is as follows:

– the number of gauging stations is proportional with the spatial variation,
– the frequency of observation depends on the temporal variation.


## 7.3  PURPOSE AND FIRST SETTING UP OF A HYDROLOGICAL NETWORK

*Purpose of a hydrological network*

The objective of a hydrological network is defined as follows:

   '*To provide for the need of hydrological information*'

The required hydrological data include water levels, discharges, rainfall and groundwater tables. The basic field data are validated and elaborated towards relevant hydrological information.

In general, two target-areas can be identified:

1. Water management, including: daily water management, alarm emergencies, set up of management models, water conservation, environmental protection etc.
2. Hydrological research, including water-balance studies, rainfall-runoff models, evaluation of effects from human interference, and specific research on ecological values.

The complete cycle of measurement, elaboration, validation, presentation and storage of data can be done either manually or fully automatically. The advantages of automatic transmission systems are:

– a clear total overview of the actual situation in the catchment area,
– reduction of costs (long term),
– ability to react efficiently upon extreme events.

*The first setting up of a hydrological network*

For the setting up of hydrological networks, the main question to be answered, reads:

> *How to design such a network providing sufficient information for the users and of minimal investments and operating costs.*

In the design procedure we may identify the following steps:
- definition of the objectives (management, research);
- to collect knowledge on the characteristics of the catchment area;
- to collect knowledge on the specific interests (agriculture, urban water management, natural environment);
- first selection of parameters to be measured, locations and frequencies;
- selection of measuring methods and equipment, as well as the selection of the transmission system and – if desired – the selection of a database management and processing system;
- structuring of the surveying organization: staff, schooling and facilities.

In this section the first selection of parameters (V), locations (L) and frequencies (f) is focused on.

The selection of the *hydrological variables* (V) is mainly governed by the objective of the network. For management purposes water levels, discharges, precipitation, and perhaps some water quality parameters should be measured. In addition for research the evapotranspiration and other parameters will be estimated.

For the selection of the most appropriate *locations* (L) the following guidelines can be hold:
- *flow measurement stations* (measurement of river flow) are desired at:
  - the outlet of catchment areas (rainfall-runoff models and water balance studies);
  - confluences and bifurcations of streams;
  - all locations where the characteristics of the catchment area or the land use is changing significantly: changes in landslope, vegetation, geological base, storage capacity and the transition between urban areas and rural areas.

  The various methods to measure river flow are discussed in Chapter 4.
- *water level gauging stations* are desired for one of the following reasons:
  - all locations where the river flow is measured;
  - locations where on-line information is necessary for flood forecasting;
  - locations where conflicting interests play a role: agriculture, water supply, environmental protection, etc.

  Both float-operated recorders, pressure transducers and acoustic sensors are applied in automatic water level gauging stations.

– *precipitation*. Information on precipitation is collected for the following applications:
  – water management, including hydrological forecasting;
  – water-balance studies;
  – studies of long-term changes of climate.
With regard to the observation-frequency distinction can be made between rain gauges (once per day) and rainfall recorders (continuous information).
Three different types of rainfall-recorders are known: tipping bucket type, syphon type and weighing type, among which the first one is most commonly used. The density of rainfall recorders in a catchment area is strongly depending on the topography of the catchment area and the type of rainfall.
– *groundwater depths* are measured for one of the following applications:
  – irrigation and drainage;
  – nature conservation;
  – withdrawal of groundwater for public water supply, industry or agriculture.

The *frequency* (f) of measuring discharges, water levels and precipitation depends mainly on the response of the hydrological processes on the precipitation and on the objectives of the desired network information. In many cases frequencies of once every 15 minutes are acceptable for water levels and discharges. Tipping bucket rainfall recorders can be adjusted to give signals (pulses) after 0.1 mm, 0.2 mm or 0.5 mm precipitation.

After a first selection of parameters (V), locations (L) and frequencies (f), the final selection is done in the following steps:
– The first selection (V, L, f) is plotted in a map of the catchment area. This is the ideal configuration of the hydrological network.
– Then a number of readjustments is carried out:
  1. Existing gauging stations: are they usable?
  2. Are the planned stations near existing cables for electricity and telephone?
  3. Is it feasible to reduce the number of flow measurement stations in case they represent catchment areas which have almost the same characteristics?
  4. Estimation of costs – investment and operation costs – may lead to a reconsideration of the planned network.
– Based upon the four readjustments the final selection of parameters (V), locations (L) and frequencies (f) will result in a *realistic hydrological network*, meeting the demands of hydrological information.

The complete procedure from measuring field data towards relevant hydrological information is presented in Table 7.3.

Table 7.3. Schematic overview of functions in a hydrological network.

| Measuring<br>– water levels<br>– discharges<br>– precipitation | ⇒ | First elabora-<br>tion<br>and recording<br>in the field-<br>station | ⇒ | Transmission<br>– manually<br>– telephone<br>– radio<br>– satellite | ⇒ | Second elaboration<br>in the central<br>station for the<br>data base |

⇓

| Information for:<br>– management<br>– research<br>– planning | ⇐ | Processing system:<br>– validation<br>– quality control<br>– statistical analysis<br>– reporting<br>– data storage |

*Network density*

The network density (area in square kilometers per station) is based on
– physiography
– climate
– population density
The World Meteorological Organization (WMO) criteria for network design are as shown in Table 7.4.

Table 7.4. Minimum density of precipitation and discharge gauging stations.

| Type of region | Area (km²) per station | |
|---|---|---|
| | Precipitation* | Flow gauging |
| 1. Flat regions of temperate mediter-<br>ranean and tropical zones | 600–900 | 1000–2500 |
| 2. Mountainous regions of temperate mediter-<br>ranean and tropical zones | 100–250 | 300–1000 |
| 3. Small mountainous islands with<br>very irregular precipitations | 25 | 140–300 |
| 4. Arid and polar zones | 1500–10.000 | 5000–20.000 |

* The precipitation figures are related to the non-recording raingauges. In those cases,
  where both non-recording and recording (automatic) raingauges are operated, the areas
  per km² for the latter type is ten times the figures in the table.

## 7.4  OPTIMIZATION OF MONITORING NETWORKS

*Why monitoring?*

In general, monitoring a system (such as a river or an irrigation area) aims at providing *information* on that system for the purpose of research or management.

If monitoring is related to research, the monitoring effort is mostly put into intensive *measurement campaigns*. These campaigns mostly serve to:
– deepening the *insight* into and improving the quantitative understanding of some specific processes within the system;
– development, calibration and verification of *mathematical models* of (parts of) the system.

Hence, these campaigns are very *problem oriented*. Once enough information has been collected, the monitoring activities usually can be stopped. Therefore, these campaigns are relatively *short-lasting* (secondary stations or stations for specific purposes).

If monitoring is related to *management*, the information usually is collected by means of routine monitoring networks.
These networks mostly have a *broad* character and are *long-lasting* (principal stations).
Routine monitoring networks depend on the *monitoring objectives*, which, in their turn, are dictated by the management objectives.

*How monitoring?*
The designer of both measurement campaigns and routine monitoring networks has to face the following problems:
– *what* is to be measured? (choice of sampling *variables, V*);
– *where* should be measured? (choice of sampling *locations, L*);
– *how often* should be measured? (choice of sampling *frequencies, f* );
– *how* should be measured? (choice and installation of measuring *equipment*).

Apart from these specific monitoring problems, he also has to solve the problem of how to *process* the *data* gathered.
This relates to the choice of:
– a suitable data base structure
– preprocessing methods
– postprocessing methods
– analysis methods

For this purposes, various agencies have developed processing systems. Such a processing system usually consists of the following components:
– data processing units
– data base
– data management system

The data processing units include data screening, data completion, data analysis, data generation and process simulation. A processing unit consists of several programs, in which each encloses an independent compilation. The integration of the programs into one system is achieved by linking them to a data base, a storage device for time series on CD Roms, magnetic discs or tapes. The data base consists of an administration file and a number of data files containing the times series.

By data management an ordering of data is pursued so that programmes have direct access to the data. It comprises the transfer of time series from table, card, tape or diskette to the data base and from data base to programme and vice-versa. Generally, the data management system controls the access to the data base.

*Why optimization?*

The most critical step in developing a successful and cost-effective monitoring program is the clear definition of information needs and monitoring objectives, which should be derived from integrated water management and policy objectives. The information needs must be clearly identified (by e.g. policy makers) and the monitoring program must respond to those information needs (quantity and quality of the information). The clear specification of monitoring objectives ensures that only necessary data is collected and that proper information is gained from the monitoring program. This promotes effective and efficient monitoring which lowers the costs of monitoring.

In practice, data from routine monitoring programs are generally used for a variety of purposes in addition to those for which the programs were designed. The costs of (routine) monitoring are generally extremely high. Important factors are the cost of equipment, installation, sample collection, sample (laboratory) analysis, processing of data, storage of data and reporting of resulting information. Considerable savings can in most cases be achieved by optimizing the monitoring network. Such optimization aims at establishing that network which provides its users with *sufficient* information on the water system against *minimal* costs. Obviously, what is 'sufficient' depends on the *objectives* to be defined by the user.

The ultimate goal of monitoring is to provide information, not data. In the past, many monitoring programs have been characterized by the 'data rich, information poor' syndrome, i.e. the focus has been primarily on the data collection aspects. There must be more attention on the analysis and further use of collected data so that the end product of monitoring is information. Data that do not contribute to identified information needs, or whose use cannot be stated explicitly, should not be collected.

CHAPTER 8

# Organization of a survey

In this chapter use has been made of the publication: 'Guidance for hydrographic and hydrometric surveys' by F.Ch. Hayes, 1978.

*General*

In order to carry out a field survey certain necessary preparations have to be made to ascertain a successful progress of the survey, and the following is essential:

- An *evaluation* has to be made *of the data required for the project on hand*, the desired *accuracy* and the *circumstances* under which the survey has to be carried out in order to decide on type and number of instruments to be used.
- If possible a *reconnaissance* has to be made of the area involved, to *select locations for the measurements* and related to this to *decide on the number of manpower*, equipment and materials, like staff gauges and triangulation beacons. If available, satellite imageries or aerial photographs of recent data should be studied, to obtain an overall picture of the survey area and its surroundings.
- An *assessment* has to be made *of all materials* which will be required during the survey, like laptops, ballpoints, pencils, wire, nails, *generators and batteries*, etc. Although among these there are small items of minor costs, they are indispensable and most probably cannot be purchased in the field.
- If frequency bound equipment is planned, like walkie talkies, information about allowed frequencies should be collected and required permits be obtained.
- *Measuring forms* have to be prepared for all sorts of measurements. To prevent loss of data, all measurement information must be recorded at least in duplicate.
- The logistical support of the survey team must be well organized to prevent discontinuity in measurements due to lack of spare parts, food or fuel.
- All instruments which are selected for the survey, *must be thoroughly checked and provided with spare parts*.

*Purpose and accuracy of measurements*

In the following list a brief review is given of the sort of measurements that have been dealt with, and their purposes.

| Sort of measurement | Purpose of measurements |
|---|---|
| Geodetic measurement | – to establish a network of beacon- and bench-marks<br>– to relate bench-marks to a reference plane by levelling |
| Positioning | – to determine the position of the survey vessel or floats during the measurements |
| Water level measurement | – to obtain water level information in open channels and subsoil<br>– to obtain correlation with discharge measurements |
| Bathymetry | – to select and determine cross sections for discharge measurements<br>– to determine dredged quantities<br>– to determine siltation and degradation |
| Velocity measurements | – to calculate discharges<br>– to obtain flow patterns in an area<br>– to obtain sediment transport data<br>– to obtain the velocity and the direction of the flow for navigation purposes |
| Discharge measurements | – to obtain rating curves (stage-discharge relations)<br>– to determine tidal volumes<br>– to determine water balances |
| Salinity measurements | – to determine salinity and temperature distribution for irrigation purposes<br>– to determine salinity intrusion |
| Sediment measurements | – to obtain data about sediment transport<br>– to obtain discharge-sediment transport relations<br>– to determine bottom composition |

Measurements have to be considered as an estimation of the measured quantity whereas the required accuracy depends on the purpose of the measurement; the absolute precision that can be obtained depends as well on the technical possibilities and limitations as on the phenomena to be measured itself.

As far as the required accuracy is concerned, no general indications can be given except that, in order to develop mathematical descriptions of certain phenomena (e.g. water movement or sediment transport) the maximum overall technical accuracy is required. This overall technical accuracy depends on:

1. The variability of the process to be measured, the knowledge hereof and the degree in which the selected survey technique is suitable to monitor the process;
2. The instrument and its calibration;

Figure 8.1. Survey boat
equipped with the Delft Bottle
Sampler (courtesy
Mr. Joop van der Pot,
WL Delft Hydraulics).

3. Experience and capability of personnel (capacity and motivation);
4. Organization and schematization of the measurements (e.g. number of verticals, sampling time, etc.);
5. Influence of the measuring device on the phenomenon itself.

In general terms, it can be said that experience and capability of personnel plays an important role in all sorts of measurements and items 1. and 3. in most of them. The influence of the measuring device on the phenomenon itself is for instance of importance when measuring bed load.

Summarizing, it is expected from a hydrographer (as far as accuracy considerations are considered) that he or she has:
– a good physical insight into the phenomena to be measured;
– a good insight into purpose and use of the measurements, and use of the data;
– a good knowledge of the properties and operational aspects of the instruments in use;
– a good knowledge of the above mentioned technical and personnel possibilities and limitations.

Only when giving all aspects their due considerations, a good, sound and useful survey can be made.

*Measuring program*
As a rule the next sequence can be followed:
– Water level recorders and staff gauges to be installed to obtain water level information from the very start of the survey;
– Bench marks, triangulation beacons and cross section marks to be established;

Figure 8.2. Small vessel for bathymetric surveys with GPS antenna (middle), reference signal for DGPS (left), and mariphone (right) (courtesy Mr. Joop van der Pot, WL Delft Hydraulics).

– Measuring and determination of the locations of triangulation beacons, marks and bench marks and the determination of heights
  of bench marks and their relation to staff gauges and water level
  recorders;
– A bathymetric survey to be carried out and/or cross sectional soundings to determine the bottom configuration of the area and to select
  locations for the various measurements;
– In tidal areas, discharge, salinity and sediment transport measurements to be planned during spring and neap tides in the wet season
  and all other measurements to be planned inbetween.
– In non-tidal areas the same measurements to be planned in the dry and
  wet season related to water levels.

When the program and the time schedule are ready, a script must be made
for the personnel involved in the measurements regarding dates and kind
of measurement, times of departures, times of anchoring in cross sections
and duration of each measurement.

The observers must be divided into groups and clear-out instructions
have to be given regarding their tasks and obligations.

It is most important to give them an outline of the purposes of the
project at hand because *nothing is more frustrating then to do a measurement without knowing its purpose.*

*Logistical support*
One of the most essential parts of the whole survey is the logistical support. If this support fails the whole survey can become a failure and all
costs and time for preparation have been in vain.

The logistical support consists of the following parts:
– food and drink supply;

Figure 8.3. Vehicle with mobile DGPS receiver and reference station for DGPS (courtesy Mr. Joop van der Pot, WL Delft Hydraulics).

- fuel supply for the vessels and speed boats;
- medical supply;
- supply of spare parts for instruments and auxiliary equipment like outboard motors, generators, etc.

Before starting the survey this support should have been fully arranged and checked so that no delay in or discontinuation of the measurements will occur.

*Measuring forms*
In order to note the measured data for each type of measurement special measuring forms must be prepared.

*Preparation of instruments*
Before starting the survey, instruments must be checked and should be overhauled and calibrated.

*Vessels*
When vessels are to be selected for the measurements, the following points should be taken into account:
- When vessels are used in rivers the draught should not exceed 1.50 m;
- Vessels must have sufficient working space preferably on the fore deck;
- Sufficient engine power to sail in high currents;
- If propellers are used, a strong davit has to be installed with sufficient height to attach the largest winch to it and with sufficient span so that the current meter is at least 1.5 metre free from the side of the vessel when the davit is swung out;
- Each vessel must be provided with a heavy anchor with a chain anchor line (rope lines to be denounced as vessels will drag their anchors during the measurements in flows exceeding 0.75 m/s);

– Cooking facilities on board;
– Sufficient accommodation so that the observers can take a rest in turns;
– Safety equipment to be on board, such as life jackets and life buoys in sufficient numbers;
– All vessels to be provided with day signals and night signals;
– Sufficient lights to be on board for night observations.

*Safety*

If measurements are to be carried out in navigation channels, local authorities should be informed in time about the locations and positions where survey vessels will be anchored to carry our their measurements during the surveying period. If feasible, radio communication should be maintained between the survey leader, the survey vessels and their base camp.

ANNEX I

# International standards

Knowledge of the behaviour of water in rivers, lakes and reservoirs is essential to the management of water systems. This knowledge is growing all the time through the efforts of hydrologists. The organization of water gauging stations is essential to efficient water management, whether in a wet country like Canada with 9 percent of all the world's fresh water or a country of deserts or ill-distributed water resources. Hydrologists have pointed out that improved forecasts for the operation of river systems have substantial economic advantages apart from their importance in measures to counter drought or flood, for example, improvements in flood inflow forecasts permit more efficient design and operation of hydroelectric power stations.

One of the most obvious uses of water resources data is in the design and operation of structures that are used to control the levels and flows of the water, particularly in the field of irrigation. The design of these structures is dependent upon technical data. Optimum design requires both accurate basic data in the project region, and sound techniques to make maximum use of those data.

ISO 'International Organization for Standardization', started its work since 1947 by providing standards of a general nature or for specific items of measuring equipment, ensuring that countries with good water management share their experience with each other and with countries where the need for this management may be a matter of life and death.

*ISO is the specialized international agency for standardization, comprising the national standards bodies of nearly 160 countries. The object of ISO is to promote the development of standards in the world with a view to facilitating international exchange of goods and services, and to developing co-operation in the sphere of intellectual, scientific, technological and economic activity.*

*ISO brings together the interests of producers, users (including consumers), governments and the scientific community, in the preparation of International Standards. ISO standards are in wide use throughout the world, in practically every area of technology, either directly or in the form of identical national standards.*

Today 69 International Standards have been prepared by ISO/TC 113, The Technical Committee for *Hydrometry*.

Most of the ISO standards have been mentioned in this book.

| Chapter | Section | Number of standards |
| --- | --- | --- |
| Water levels | 2.9 | 6 |
| Measurement of bed levels | 3.9 | 3 |
| Discharge measurements | 4.13 | 34 |
| Measurement of sediment transport | 5.9 | 7 |
| Flow measurement structures | 6.13 | 16 |

The following three standards have not yet been mentioned in the sections:

ISO 772      Vocabulalry and symbols
TS 25377     Hydrometric Uncertainty Guidance  (HUG)
CEN 13798    Specification for a reference raingauge pit

The large majority of the standards have been drawn up by ISO/TC 113. At the European level 30 national european standards bodies are participating in the CEN (Comité Européan de Normalisation), the European committee for standardization. CEN/TC 318 is the technical commitee for Hydrometry. ISO/TC 113 and CEN/TC 318 are cooperating in the Vienna Agreement.

ISO Central Secretariat
P.O. Box 56
CH - 1211 Geneva 20
Switzerland
    Email: central@iso.org
    Web:   www.iso.org

Secretariat ISO/TC 113
Bureau of Indian Standards, B.S.Z. Marg,
New Delhi - 110002 India

Secretariat CEN/TC 318
BSI Group Headquarters
389 Chiswick High Road, London
W4 4AL England

Dutch Shadow Committee Hydrometry (NEN)
P.O. Box 5059
2600 GB Delft, The Netherlands

# References

Ackers, P., White, W.R., Perkins, J.A. & Harrison, A.J.M. 1980, *Weirs & Flumes for Flow Measurement*. John Wiley & Sons.

Aisenbrey, A.J. 1974, *Design of small structures*. US Dept. of the Interior, Bureau of Reclamation, Denver Colorado.

Alsthom, Fluides, Neyrtec Distributions, Amil-, Avio- & Avis Gates, Services Techniques et Commerciaux, La Courneuve, France.

Askew, A.J. 1989, *Network Design*. Internat, Inst. for Hydraulic & Environmental Engineering (IHE), Geneva/Delft.

Bodhaine, G.L. 1969, *Measurement of Peak Discharge at Culverts by indirect methods*. US Department of the Interior, Book 3, Chapter A3. Washington.

Boiten, W. 1985, *The Rossum weir*. Polyt. Tijdschrift.

Boiten, W. 1987, *Hobrad weirs*. Polytechnisch Tijdschrift pt/c (41)1.

Boiten, W. 1990, Hydraulic design of the pool-type fishway with V-shaped overfalls. In: *Proceedings of the International Symposium on Fishways '90* in Gifu, Japan.

Boiten, W. 1990, *Discharge relation of the pool-type fishway with V-shaped overfalls*. Wageningen University, Section Water Resources, report 6 (in Dutch).

Boiten, W. June 1992, *Vertical gates for the distribution of irrigation water*. WL Delft Hydraulics publication 465.

Boiten, W. April 1993, *Training course on Discharge Measurement Techniques for the project: Sources for Sana'a Water Supply*, WL Delft Hydraulics, Q 1733.

Boiten, W., Dommerholt, A. & Soet, M. 1995, *Handboek debietmeten in open waterlopen*. Wageningen Agricultural University, Department of Water Resources, report 51 (in Dutch).

Boiten, W. 1996, Flow measurement in open channels for hydrological networks. In: *Utilización y manejo de los recursos hídricos*, UNA, Heredia, Costa Rica.

Boiten, W. 2002, Flow measurement structures. In: Flow Measurement and Instrumentation 13. pp 203–207.

Boiten, W. & Dommerholt, A. 2006, Standard design of the Dutch pool and orifice fishway. In: *Intern. Journal River Basin Management (J.R.B.M)*. volume 4, no. 3, 2006.

Bos, M.G., Replogle, J.A. & Clemmens, A.J. 1984, *Flow Measuring Flumes for Open Channel Systems*. John Wiley & Sons.

Bos, M.G. (ed.) 1989, *Discharge Measurement Structures*. International Institute for Land Reclamation & Improvement/ILRI, Wageningen, The Netherlands. Third edition.

Bos, R.J. 1961, *The Romijn measuring gate*. Delft Hydraulics.

Butcher, A.D. 1921, *Discharge over clear overfall weirs*. Delta Barrage, Cairo.

Day, T.J. 1977, Observed mixing lengths in mountain streams. *Journal of Hydrology* 35: 125–136.

Dommerholt, A. 1992, *Chimney effect in measured waterlevels*. Wageningen University, Section Water Resources, report 29 (in Dutch).

Doorenbos, J. 1976, Agro- meteorological field stations. FAO report 27, Rome.

Durst, F. 1987, *Discharge measuring methods in pipes*. Institut für Hydromechanik und Wasserwirt- schaft ETH, Zürich.

Edwards, Th.K. & Douglas Glysson, G. 2005, Field methods for Measurement of Fluvial Sediment. USGS.

Endress & Hauser, B.V. Naarden, The Netherlands.

Fieldservice Technical Paper 001, October 1996. Broad Band ADCP. Advanced Principles of Operation. R.D. Instruments. San Diego, California, USA.

Franke, P.G. 1974, *Abflusz über Wehre und Überfälle*. Bauverlag Wiesbaden und Berlin.

Furness, Richard A, *Fluid Flow Measurement*, Longman in association with The Institute of Measurement & Control, UK.

Gaeuman, D. & Jacobson, R.B. 2005, Aquatic Habitat Mapping with an Acoustic Doppler Current Profiler: Considerations for Data Quality. USGS, Open-File Report 1163.

Gasser, M.M. 1991, *Measurements of Bed Load Transport in the Nile River at Bani-Mazar with the Delft-Nile Sampler*. Hydraulics & Sediment Research Institute (HSRI), Delta Barrage, Cairo.

Gordon, R.L. January 1996, *Acoustic Doppler Current Profiler*. Principles of Operation. R.D. Instruments. San Diego, California, USA.

Grant, D.M. 1985, *Open channel flow measurement handbook*. Isco Inc. Lincoln, Nebraska.

*Guidebook to Doppler Flow Measurement in Liquids*, 1989, Polysonics, Houston Texas.

Hager, W.H. 1985, Der Mobile Venturikanal, *Gas-Wasser-Abwasser* nr. 11.

Hager, W.H. 1983, Modified Venturi Channel. *Journal of Irrigation & Drainage Engineering*.

Hager, W.H. 1986, Modified Trapezoidal Venturi Channel. *Journal of Irrigation & Drainage Engineering*.

Hager, W.H. 1987, Lateral Outflow over Side Weirs. *Journal of Hydraulic Engineering*, Volume 113.

Hager, W.H. 1988, Mobile Flume for Circular Channel. *Journal of Irrigation & Drainage Engineering*.

Hager, W.H. 1993, Abflusz über Zylinderwehr. *Wasser und Boden*.

Haldar, S.K. 1981, Problems of stream flow measurement in hilly regions. *Irrigation & Power* 38(1981): 227–237.

Hayes, F.Ch. 1978, *Guidance for hydrographic & hydrometric surveys.* Delft Hydraulics, Publication 200.

Herschy, R.W. 1978, *New Technology in Hydrometry.* Adam Hilger Ltd., Bristol & Boston.

Herschy, R.W. (ed.) 1978, *Hydrometry: Principles & Practices.* John Wiley & Sons.

Herschy, R.W. 1985, *Streamflow Measurement.* Elsevier Applied Science Publishers, London.

Herschy, R.W. 2002, The world's maximum observed floods. In: Flow Measurement and Instrumentation 13. pp 231–235.

Huisman, P. 2004, Water in The Netherlands. Hydrological Society (NHV).

Hurn, J. 1989, *G.P.S. A Guide to the Next Utility.* Trimble Navigation Limited.

Hurn, J. 1993, *Differential GPS Explained.* Trimble Navigation Limited.

*Hydrologists Field Manual*, 1988, Publication 15 of the Hydrology Centre, Christchurch, New Zealand.

*Instromet, Flow-2000*, An acoustic flowmeter system. Instromet Ultrasonic Technologies, Silvolde, The Netherlands.

International Hydrographic Organization. 2005, Manual on Hydrography International Hydrographic Bureau, Publication M-13.

International Standards Organization (ISO); All the International Standards on flow measurements in open channels. ISO Central Secretariat, Case Postale 56, CH-1211 Genéve 20, Switzerland.

Jansen, P. Ph. et al. 1979, *Principles of River Engineering*, Pitman, London.

Jansen Venneboer, Vijzelpompen, Industrieweg 4, Wijhe.

Jarett, R.D., Hydraulics in Mountain Rivers. In: *Channel Flow Resistance, Centennial of Manning's Formula*, (Ben Chie Yen, ed).

Kay, M. & Hatcho, N. 1992, *Small-scale pumped irrigation, energy & cost.* Food & Agriculture Organization (FAO), Rome.

Kilpatrick, F.A. & Schneider, V.R. 1983, *Use of flumes in measuring discharge.* Book 2, Chapter A14. United States Geological Survey.

Kolupaila, S. 1961, *Bibliography of Hydrometry*, Notre Dame, Indiana, Univ. of Notre Dame Press, 975 pp.

Kraijenhoff van de Leur, D.A. 1972, Hydraulica 1, Lecture Notes, Wageningen Agricultural University.

Lauterjung, H. & Schmidt, G. 1989, *Planning of Intake Structures*, Friedr. Vieweg & Sohn, Braunschweig/Wiesbaden.

Linsley, R.K. & Franzini, J.B. 1979, *Water Resources Engineering*, McGraw-Hill Book Company.

Loedeman, J.H. July 1993, A Useful Tool in Training Non-professional Land Surveyors in GIM.

Loedeman, J.H. et al. 1998, Lecture Notes GIS. Wageningen Agricultural University.

Made, J.W. van der (ed.) 1986, *Design aspects of hydrological networks.* TNO Committee on hydrological Research. The Hague.

Magalhães, A.P. & Lorena, 1989, *Hydraulic Design of Labyrinth weirs.* Laboratorio Nacional de Engenharia Civil, Lisboa.

Majewski, W., Baginska, M. & Walczak, P. April 22–24, 1985, Determination of roughness coefficients for ice-covered rivers by means of direct measurements of velocity distribution. *IAHR, Proceedings Symposium Measuring techniques in hydraulic research*, Delft, The Netherlands.

*MCB Shaftencoders* (in dutch), Van Reysen, P.O. Box 5005, 2600 GA Delft.

Melvyn, Kay 1986, *Surface Irrigation, Systems & Practice*, Cranfield Press, Cranfield.

Meyer, D.G. 1992, *Dilution method for measurements of unsteady discharges in mountain streams*. Delft University of Technology, Civil Eng. Faculty.

Morlock, S. et al. 2002, Feasibility of Acoustic Doppler Velocity meters for the production of Discharge Records from the USGS Streamflow-Gaging Stations. USGS Indianapolis, Water Resources Investigations Report 01-4157.

Muysken, J. 1932, Berekening van het nuttige effect van de vijzel, *De Ingenieur* (in dutch).

Nedeco, 1959, *River Studies & Recommendations on Improvement of Niger & Benue*. North-Holland Publ. Amsterdam.

Nedeco, 1965, *Siltation*, Bangkok Port Channel, Nedeco, The Hague.

Nedeco, 1973, Rio Magdalena & Canal del Dique Survey Project. Nedeco.

Normann, J.M. et al. 1985, *Hydraulic design of Highway Culverts*. US Department of Transportation.

Ogink. H.J.M., *HYMOS, a date processing system for hydrological time series*. Delft Hydraulics.

Ott Messtechnik GmbH, Kempten, Germany.

Ott Hydrometry, Thalimedes float operated shaft encoder, with integrated data logger. Ott Messtechnik, Postfach 2140, D-87411 Kempten, Germany.

Polman, J. en Salzman, M.A. 1996, *Handleiding voor de technische werkzaamheden van het kadaster*. Kadaster Apeldoorn (in dutch).

Riggs, H.C. May–June 1976, A simplified slope-area method for estimating flooddischarges in natural channels. *Journ. Research US Geol. Survey*, 4(3).

Rijn, L.C. van. 1986, *Manual Sediment Transport Measurements*. Delft Hydraulics, March.

Seba Hydrometry GmbH, Kaufbeuren-Neugablonz, Germany.

Shaw, Elizabeth M. 1988, *Hydrology in Practice*, VNR International, second edition.

Simpson, M.R. 2001, Discharge Measurements using a Broad-Band Acoustic Doppler Current Profiler. USGS, Sacramento, Open-File Report 01-1.

Spaans Babcock bv, *Vijzelpompen*, Spaans, Hoofddorp.

Sturm, T.W. 2001, Open Channel Hydraulics. Georgia Institute of Technology.

Teunissen, P.J.G. & Kleusberg, A. 1998, *GPS for Geodesy*, 2nd Edition, Springer.

Thysse, J.Th. 1951, Hydraulica. In: *De Technische vraagbaak* (in dutch).

Tilrem, O.A. 1986, Methods of measurement & estimation of discharges at hydraulic structures. Operational Hydrology Report No. 26, WMO-No. 658, World Meteorological Organization, Geneva.

Topcon Optical Co. Ltd., Tokyo, Rangefinder, Model DM-50U.

Trimble Navigation Limited, 1996, Mapping Systems, General Reference.

UNESCO IHP-IV Project E-1.2, 1994, *Applied Hydrology for technicians*. UNESCO, Paris.

US Army Corps of Engineers. 2002, Engineering and Design Hydrographic Surveying. USACE Manual, no. 1110-2-1003.

Ven Te Chow 1959, *Open Channel Hydraulics*, Mc Graw-Hill Book Company, London.

Verhoeven, R. 1989, *Hydrometrie*. Rijksuniversiteit Gent (in Dutch).

Water Survey of Canada, Hydrometric Field Manual. Levelling, 1984. Measurement of Stage, 1983. Stream-gauging cableways, 1984. Measurement of stream flow, 1981. Automated Moving Boat System, 1985. Inland Waters Directorate, Ottawa, Canada.

White, W.R. 1975, *Thin plate weirs*. Hydraulics Reseach Station, Wallingford.

WMO Commission for Hydrology, 1971. Use of weirs & flumes in stream gauging. WMO-no. 280, Geneva.

WMO (World Meteorological Organization), 1980. Manual on stream gauging. Volume 1: Fieldwork; Volume II: Computation of discharge. Operational Hydrology report no. 13. WMO-Geneva, Switzerland.

WMO, 1981, Guide to Hydrological Practices, Volume 1, Data acquisition & Processing, WMO-no. 168, Geneva.

WMO, 1994, Applied hydrology for technicians, Volume III, IHP-IV, Project E-1.2, UNESCO, Paris.

WMO, 1996, The adequacy of hydrological networks: a global assessment.

WMO, 1996, Intercomparison of Principal Hydrometric Instruments, third phase. Data Telemetry & Transmission systems, WMO.

WMO, 1996, Integrated Hydrological Networks, WMO.